Visual Art as Theology

NEW STUDIES IN AESTHETICS

Robert Ginsberg
General Editor

Vol. 21

PETER LANG
New York • Washington, D.C./Baltimore • San Francisco
Bern • Frankfurt am Main • Berlin • Vienna • Paris

Barbara Dee Bennett Baumgarten

Visual Art as Theology

PETER LANG
New York • Washington, D.C./Baltimore • San Francisco
Bern • Frankfurt am Main • Berlin • Vienna • Paris

Library of Congress Cataloging-in-Publication Data

Baumgarten, Barbara Dee Bennett.
Visual art as theology / Barbara Dee Bennett Baumgarten.
p. cm. — (New studies in aesthetics; vol. 21)
Includes bibliographical references and index.
1. Aesthetics—Religious aspects. 2. Art and religion. 3. Image (Theology) 4. Theology—20th century. 5. Polanyi, Michael, 1891–
6. Tillich, Paul, 1886–1965. I. Title. II. Series.
BL65.A4B38 1994 291.3′.7—dc20 93-46854
ISBN 0-8204-2462-5 CIP
ISSN 0893-6005

Die Deutsche Bibliothek-CIP-Einheitsaufnahme

Baumgarten, Barbara Dee Bennett:
Visual art as theology / Barbara Dee Bennett Baumgarten. - New York; Washington, D.C./Baltimore; San Francisco; Bern; Frankfurt am Main; Berlin; Vienna; Paris: Lang, 1994
(New studies in aesthetics; Vol. 21)
ISBN 0-8204-2462-5
NE: GT

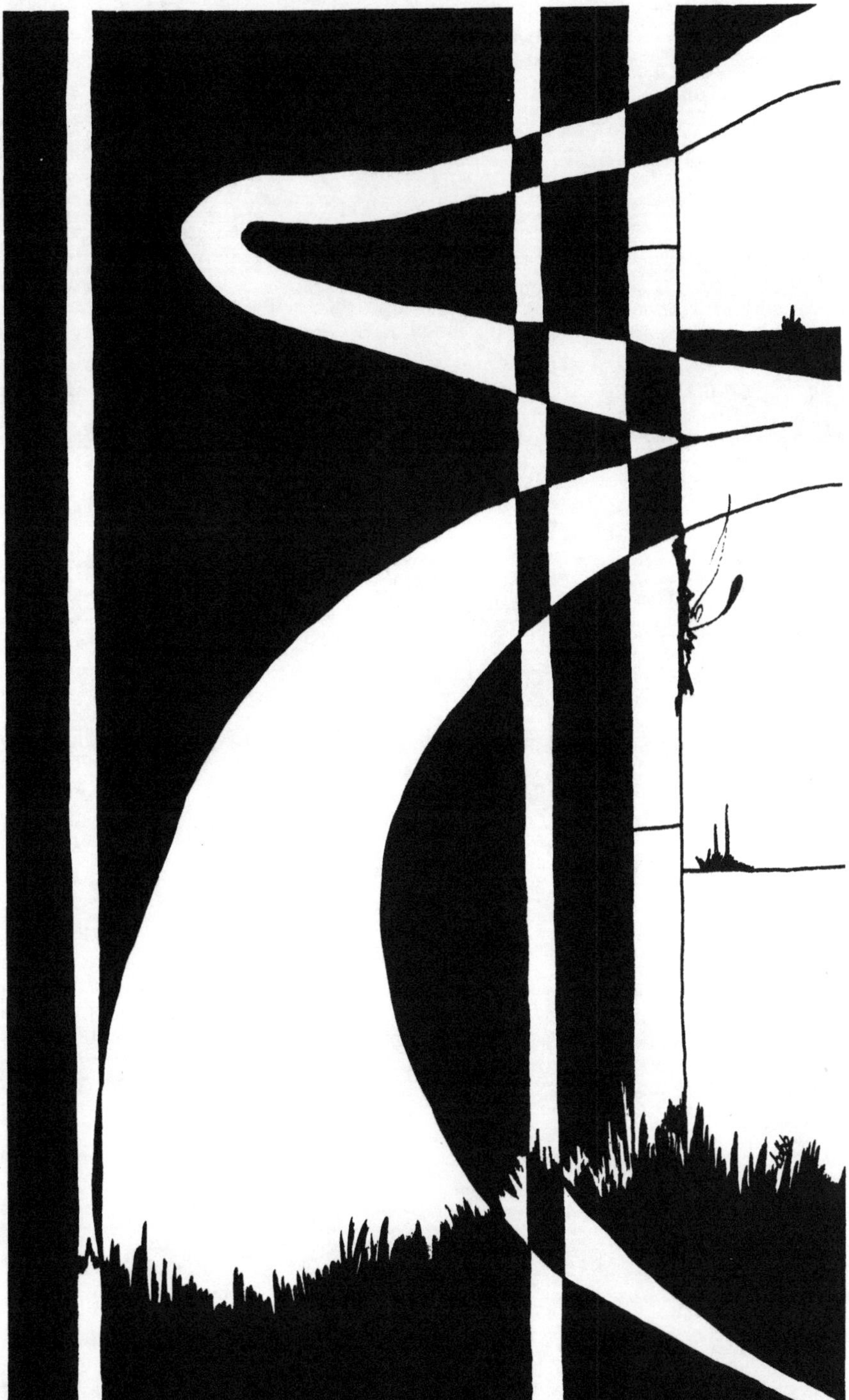

To Bill, Jessica, Austin, and Bennett

Contents

List of Abbreviations

CI "The Creative Imagination"

KB *Knowing and Being: Essays*

M *Meaning*

M:P "Meaning: A Project"

OAA *On Art and Architecture*

PK *Personal Knowledge*

SFS *Science, Faith, and Society*

SM *The Study of Man*

TC *Theology of Culture*

TD *The Tacit Dimension*

WP "What Is a Painting?"

WR "What Is Religion?"

List of Illustrations

The author and publisher thank the custodians of the following works of art for supplying photographs and granting permission to use them.

1. B. Dee B. Baumgarten, *World*, 1986, Private Collection.
2. B. Dee B. Baumgarten, *Dance of the Twigs*, 1976.
3. W.A.K. Johnston, *The Bones* (1857; William Turner and John Goodsir, *Atlas of Human Anatomy and Physiology* (Plate 1). Courtesy of the National Library of Medicine, Washington.
4. George Segal, *Walk–Don't Walk*, 1976. Collection of Whitney Museum of American Art, New York. Purchase, with funds from the Louis and Bessie Adler Foundation, Inc., Seymour M. Klein, President, the Gilman Foundation, Inc., the Howard and Jean Lipman Foundation, Inc., and the National Endowment for the Arts.
5. Sandro Botticelli, *Madonna and Child with Singing Angels* (ca. 1477; Gemaeldegalerie, Staatliche Museen, Berlin, Germany). Foto Marburg/Art Resource, NY.
6. Paul Cézanne, *Madame Cézanne in a Red Dress*, 1839-1906. The Metropolitan Museum of Art, New York. Purchase, Mr. and Mrs. Henry Ittleson Jr. Fund, 1962.
7. Christo, *The Umbrellas, Joint Project for Japan and U.S.A.* (1990, The Tejon Pass, U.S.A.). Photograph by the author.
8. Giotto di Bondone, Scene from Life of St. Francis fresco cycle: *St. Francis Renouncing His Father* (ca. 1296-1300; Bardi Chapel, S. Croce, Florence, Italy). Alinari/Art Resource, NY.
9. Pablo Picasso, *Guernica* (1937; [Copyright 1994 Artists Rights Society, New York/SPADEM, Paris] Centro de Arte Reina Sofia, Madrid, Spain). Giraudon/Art Resource, New York.
10. Jan Steen, *The Dancing Couple*, 1663. The Widener Collection, National Gallery of Art, Washington.

Preface

I am a visual artist. I draw, I paint, I quilt. In 1984, I entered seminary to prepare myself for ministry as a fine artist. I wanted to understand the tradition of the visual arts in the church: its symbols and its impact on theology. Paul Tillich's informal aesthetics were helpful and problematic to my inquiry. I read Tillich to sort out my own ideas. I found his theology to be helpful but his ideas on art limited.

My art grew tremendously during that time of academic study. But it was not enough. My thirst to understand the relationship between the visual arts and theology was merely whetted. The doctoral program at the Graduate Theological Union which concentrates on the arts and religion proved to be a lure I could not resist.

Between my seminary work and doctoral program, I discovered that many texts on art and religion refer to Alfred North Whitehead. Thus, I began my doctoral program with the course, "Ethics in Process: a study in the thought of Michael Polanyi and Alfred North Whitehead," co-taught by Benjamin Reist from the San Francisco Theological Seminary and Charles McCoy from the Pacific School of Religion. I was not aware of Michael Polanyi's thought, but I had a hunch I needed to know about Whitehead's ideas. As the course progressed, I became enchanted with Polanyi. It was his epistemology, I ascertained, that had the potential to elucidate the relationship between art, religion, and theology. And Polanyi helped me to better understand Tillich.

Writing about art is risky business. Art is essentially indeterminate and ambiguous. The power of art lies in its expression of a range of meanings, including ambiguous or contradictory meanings, by visual means. Art cannot be fully described, only experienced.

The intent of aesthetics is to point beyond itself to an enhanced, but not descriptive, understanding of art. Language may limit and cannot determine our understanding of art. Aesthetics summons language into an articulation of what cannot be articulated. This limitation causes my language to be at times awkward and confusing. I have attempted to write clearly and simply, to know and respect my verbal limits. But the boundary between the articulate and non-articulate is vague. I ask the reader to bear with my language when it appears confusing or unduly complex, for I am striving to write within the indeterminate range between language's limit and art's reality.

The following text is the result of my attempt to sort out a theological dimension of art. My heartfelt thanks to those who have helped me along the way, especially, Charles McCoy's encouragement in my learning process; Doug Adams's creative enthusiasm which kept me going; John Dillenberger's expertise in Tillich's aesthetics; Jane Dillenberger's art critical perspective which was essential; and Phil Mullins's insights into Polanyi's epistemology and the arts which often cleared my thoughts on the matter. Finally, my gratitude goes to Wendy Larner for her critical reading and creative response to my text.

My gratitude goes to my companions at Peter Lang Publishing: Heidi Burns, for her care and support; Robert Ginsberg, for his editing expertise and wonderful humor; and Christine Marra and Nona Reuter, for their skill, patience, and good work.

The publisher and I thank for permission to reprint quotations: Thomas A. Langford and William H. Poteat, eds. *Intellect and Hope: Essays in the Thought of Michael Polanyi*. Durham, N.C.: Duke University Press, 1968. Reprinted by permission of the publisher. All rights reserved. Michael Polanyi, *Personal Knowledge*. Chicago: University of Chicago Press, 1958; corrected ed., 1962. © copyright 1958, 1962 by Michael Polanyi. Reprinted by permission of the publisher. All rights reserved. Paul Tillich, *On Art and Architecture*. John and Jane Dillenberger, eds. New York: Crossroad Publishing Co., 1987. Reprinted by permission of the publisher. All rights reserved.

Barbara Baumgarten

Santa Paula, California
April 1994

PART I:

MAKING AND VIEWING ART

AND

THE EPISTEMOLOGY OF MICHAEL POLANYI

About Michael Polanyi

Michael Polanyi was born in Hungary in 1891 to a wealthy Jewish family infused with intellectual and artistic interests. During his youth, Polanyi's family lost much of their wealth, and Michael reluctantly chose a medical career in the Austro-Hungarian army during World War I. Jewish heritage and impoverishment persuaded Polanyi to resist attempting University appointment in his chosen field of chemistry. His interest in chemistry was too strong to forego, however, and after the war he abandoned medicine to begin a career as a physical chemist at the Kaiser Wilhelm Institute in Berlin. With the rise of Hitler, Polanyi found it necessary to leave Germany. With his wife and two sons, he journeyed to England where he accepted a post as Professor of Physical Chemistry at the University of Manchester. After many successful (and creative) years in this post, Polanyi chose to change careers again. He became Professor of Social Studies at Manchester and began publishing in economics and philosophy.

Polanyi, the philosopher, pioneered a post-critical epistemology that was held in suspicion by both rationalist and anti-rationalist philosophical camps. He did not adhere to the established ideologies that rely solely on intellect nor to proposals that gave priority to emotion. Polanyi was primarily concerned with the relation of science to society and the personal involvement of the scientist in his/her pursuit of discovery. In working out this concern, Polanyi provides a balance between reason and emotion curtailed by modernism.

Upon retiring from the University of Manchester, Polanyi worked for two years as a Research Fellow at Merton College at Oxford. Following an illustrious and prolific career, he died at Oxford in 1976.

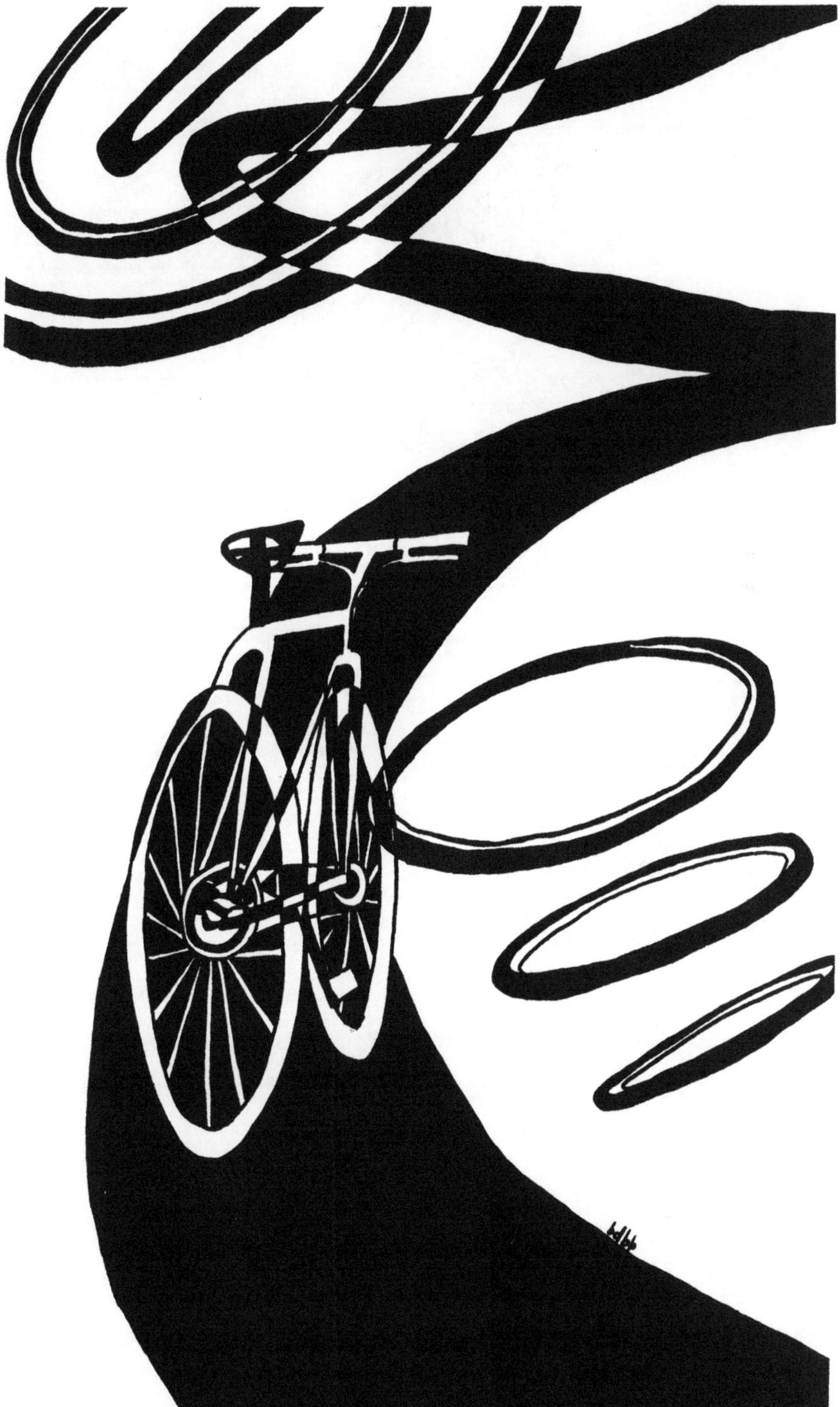

Chapter One

Encounter In Our World

About ten o'clock one Sunday morning in January, I formally placed myself in a kind of laboratory. The laboratory was principally of my own making, and consisted, insofar as its physical attributes were concerned, of two chairs, one at a table at which I would write, the other at a large slant-top desk at which I would paint...This part of the laboratory, this workroom, was about thirty kilometers from Mexico City... Certain things were clear to me at the outset of work on that Sunday morning. It was clear, for instance, that I would start writing a book and that this book would try to tell a story of self-discovery.[1]

*****

To discover new forms, new symbols, new patterns, in short, to create, requires courage. That courage is to be fully committed, yet aware at the same time that we might be wrong. Often immersion into chaos occurs in order to bring forth form. Creation demands a push into the unknown, the new. We are never satisfied with what we know now; truth lures us further into its depths. Our knowledge of reality—truth—is always partial. Creativity beckons us to discover beyond what has become conventional; it is an ever-moving probe into depths unrealized. What is the nature of this creative drive, unique yet foundational to being human?

Creativity is the process of bringing something new into being. It is not a hobby but demands skill and commitment. It is not passive but active. It is an encounter with ourselves, our culture, our world. As our encounter deepens so does our creativity. Insight occurs through indwelling. The question is often posed, "why does an original idea 'pop up' unexpectedly?" It doesn't, unless we have been absorbed in exploring something. Rollo May describes this in his book *The Courage to Create*:

Insights pertain to those areas in which the person consciously has worked laboriously and with dedication.... We cannot will creativity. But we can will

> to give ourselves to the encounter with intensity of dedication and commitment. The deeper aspects of awareness are activated to the extent that the person is committed to the encounter.[2]

Michael Polanyi uses the term "heuristic" to describe the process of discovery. While striving for insight, the seeker strains in search of clues which offer a possible solution, often in a seemingly spontaneous manner. The "personal form of knowing" that Polanyi emphasizes is rooted in our commitments, passions, and judgments. We rely on these through our embodied selves in the world to know. Heuristics overcomes a subjective reading of personal knowing because it excludes fanciful endeavors that are not feasible, or that are at least not in the realm of feasibility. Reality makes claims upon and guides our pursuits to know. The knower responds to reality and its range of possible meanings in a striving manner to understand what s/he does not yet grasp, the as yet-unknown. Discovery is not determined or spontaneous, but is due to the intentional effort of a person to comprehend.

In addition, knowing changes us. The successful solution of a problem evokes new capacities within us. When the sought-for is comprehended, we commit to a new view of reality. We integrate our discoveries into our selves so that we become new beings. All the while, who we have become (are becoming) is ever continuous with and reliant upon who we were. These elements, (1) the striving for the as yet-unknown, (2) the lighting upon its solution, and (3) our being changed by the solution, are what comprise our heuristic powers of creativity.

Creativity is the process of bringing something new into being *through* the skilled, active, and committed encounter of the individual with his or her world. "Creativity" is a comprehensive term that includes learning and new knowledge of all types. I use artistic creativity and the making of art to exemplify the post-critical epistemology of Michael Polanyi.

Though Polanyi's work concentrates on a philosophy of science, it has far-reaching implications for all manner of knowing. His consideration of art is brief but dense.[3] Polanyi's writings which explicitly involve themselves with art are valuable to my inquiry, but so too are his writings on science. The epistemology Polanyi develops based on scientific discovery can be extended to include an epistemological understanding of the visual arts. He has concluded, "all knowledge is of the same kind as the knowledge of a problem."[4]

The thought of Polanyi sheds light on art and creativity, and this inquiry illuminates Polanyi's epistemology. Polanyi's view and the role of creativity may be mapped by the perspective of the artist's experience of making art. Donald Weismann's narrative, "In the Pursuit of Discovery," and my personal training in art confirm that Polanyi's epistemological schema involves artistic as well as scientific discovery. In discovery, the seeker begins with merely a hunch to where s/he is going, while the exact route is imprecise; the seeker embarks on a path, trusting in clues that point to an as yet-unknown aspect of reality.

The foundational set of clues upon which we rely to know reality includes our particular, historical, and social location. A network of tacit knowing functions as an active component of our discovery of reality. This labyrinth of tacit knowing can be more clearly understood by inquiring into our world and our encounter with it.

1. World

As finite beings, we are located within an historical and social context that affects the lens of our experiencing reality. We are born on a certain day, in a certain place, among a particular group of people. This is our world. As we grow, our world expands to include more time, more places, more people. It becomes complex and changing while forming a tangled wholeness that can only be partially specified. Charles McCoy names this particularity of our world "location." He identifies some of the dimensions of human location that are distinct enough to specify.

First, we share space with others. To live is to act and interact among others in our homes, work, play. Communal, local, national, and international space is shared with us. Second, because of inhabited space, we live in an atmosphere of wariness and watchfulness. We cannot know what others will do. Life with others is always in principle surprising and demands that we live with caution and also with risks. Third, human location is grounded in time: a moving present is connected with a past and a future. Linear time is an element of a dynamic time brimming with all that we are, have been, and will be.

Fourth, we are a particular person sharing with but different from all others. The specifics of our particularity may be close to another person, as with a family member, or vastly different, as with someone from another culture. Through our particularity, we become aware of the plurality of people, experiences, events, yet at the same time we recognize commonality. We are not completely limited to our particularity but can always expand it through dwelling in accounts of the experiences and ideas and ideas of others. This is primarily achieved through the arts and humanities. The arts invite us to leave our present location to that presented in the artwork. We can leave the fields of Kansas to stand in the streets of Dante's Florence or the countryside of Pissarro's France. We can live, through the arts and humanities, in multiple locations.

Fifth, the location in which we dwell as human beings has relations and events woven together into a history. This history, as the story of our lives, gives significance to who we are. The particulars are integrated into a whole that yields meaning. Sixth, we are involved in being human. We actively dwell in our world, committed to becoming all that we are. Our identity and existence derive from and are invested in our world.

Finally, we live toward a reality that defines our location and provides a perspective from which we operate as human beings. Thus, we could speak of ourselves as "emerging and acquiring self-understanding in community and of that community being given pattern and identity by its historical consciousness and by the believed-in deity that shapes its action."[5]

These dimensions of location are the beginnings of an understanding of our world. Ultimately, our world is mysterious, and largely unspecifiable; we can never achieve an absolute view of our world, nor of anyone else's. We dwell in our world which functions tacitly in our achievement of skills which guide our pursuit of reality.

Our world is always a part of who we are; we can never wholly leave it for another. If I choose to become a Buddhist after being raised in a Western Christian location, I would be unable to be a Buddhist like one whose only location had been Buddhist. After years in my new perspective, I would still tacitly hold my original location. Thus, my experience as a Buddhist would vary from the person without the former Christian exposure. Our world is a place of meetings, the Western Christian in me meets the Buddhist in me to create a new being, a new location.

The artist encounters this world in the process of creation. Donald Weismann set up a laboratory in Mexico to inspire his creative pursuit. He removed himself from his familiar world of Milwaukee to the unfamiliar world of a Mexican countryside thirty kilometers outside of Mexico City. As Weismann withdrew himself from community and dwelt in the unfamiliar space of the plain room he had set up for himself, he found his actual world intensifying.

A saying is that absence makes the heart grow fond. While an emotional response of fondness was not necessarily Weismann's experience, an acute awareness of his world came into focus:

> ...I had the feeling of the presence of my entire life, replete, in some way, with the cities I had known, the intimate details of the rooms in the houses where I had lived, the colors and textures of the streets I had walked, the seasons I had felt against my skin—their blending from hot to cold and from rain to snow; the people—all of them—their gaits, their clothes, the way they held their heads and hands, the sounds of their voices and the implications of their conversations; the thousands of skies, millions of sounds and sights, textures and smells, and the billions of transmutations of these events and feelings that were my life.[6]

Described in Weismann's narrative is the location of inhabitation, dynamic time, particularity, history, and involvement. What usually functions tacitly took on an intense focal awareness. Weismann had a hunch that the key to the creation of a story lay inside the laboratory within himself. Discovering meaning meant that he needed to grapple with the substance of his life experience. For Weismann, the key to the creative experience was an encounter with his world.

2. Encounter: The Tacit Dimension

From the moment we are born, learning takes place through encounter with our world. We are born into a social context filled with antecedent relationships. We gain selfhood within our given social and relational context. Our earliest relationships to our world are primarily tactile. Our initial experiences that stimulate learning are through touch: being touched and embraced and reaching out to touch and handle ourselves and all that is around us. This touching, stimulation, and our active response is essential to human growth. Without touch, the infant wastes away and dies. Bodily tactile experience, an active engagement with our world, is foundational to human knowing and being.

Knowing is personal and active. From infancy our emergence as "knower" is based on our personal and active engagement with our bodies, the people around us, all that surrounds us. Learning is never passive but is an active process of discovery. The infant discovers it has fingers that can grasp, a cry that brings response, a body separate from its mother. For Polanyi it is this, discovery, that is knowledge's most significant form.

Discovery is made through an active inquiry in which we strive to know reality's intrinsic intelligibility. Discovery is the acquisition of new knowledge which could not be inferred from what we already know.

Discovery is inexact, for it is a skill and an art which relies on unspecifiable clues and personal judgment rather than some contrived exacting system. While it may not be precisely explained, discovery is not arbitrary or accidental. This is key in the arts, especially abstract expressionism.

Many believe that art is often begotten through a happy accident or is accidental altogether, as has been suggested regarding Jackson

Pollock's drip paintings. Yet the ability to create a work of art does not rely on some kind of gambler's luck. It depends on talent, nurtured and guided by discipline, instruction, and attentive endeavor. Painting, like all knowing in Polanyi's epistemology, is based on the fact that *we can know more than we can tell.*

A painter presents a completed painting to a viewer. The viewer asks the painter how a specific color was achieved. The painter hems and haws, offers a few general ideas of color mixing that were probably used based on the painter's knowledge of color theory, but is unable to tell exactly how much of which colors were used in the mix. The color achievement is non-specifiable. The painter knows how to make the color but cannot explicate a recipe for exact reproduction.

One of the most difficult of the artistic tasks is knowing when a painting is complete. Overpainting is a common problem of the novice. However, the painter cannot describe why s/he knows that the painting is complete. A novice learning to paint acquires this skill through looking at the masters, internship with an experienced painter, and practice. Knowledge of when a painting is complete is an indefinable skill that demands indwelling before achievement is possible.

Michael Polanyi describes the process in the following passage in terms of science. Since what he states is true also for the creation of art, I have substituted the terms appropriate to art in brackets where he mentions science.

> Since an art cannot be precisely defined, it can be transmitted only by examples of the practice which embodies it. He [*sic*] who would learn from a master by watching him must trust his example. He must recognize as authoritative the art which he wishes to learn and those of whom he would learn it. Unless he presumes that the substance and method of [art] are fundamentally sound, he will never develop a sense of [artistic] value and acquire the skill of [artistic] enquiry. This is the way of acquiring knowledge, which the Christian Church Fathers described as *fides quaerens intellectum*, 'to believe in order to know.' To learn an art by the example of its practice is to accept an artistic tradition and to become a representative of it.[7]

To use an example of Polanyi's, we recognize a particular face in a crowd, but we cannot describe in words how we know that face. We know the face, just as the painter knows the color, the completed painting, through tacit knowing.

Knowing incorporates two kinds of awareness: the parts, details, particulars *from* which we attend, and the whole or meaning *to* which we attend. This is the characteristic structure of tacit knowing. But Polanyi speaks of these two things in different ways. In his earlier writings he describes tacit knowing as consisting of two kinds of awareness: the subsidiary and the focal. Later, in *The Tacit Dimension*, he describes tacit knowing as consisting of two terms: the proximal and the distal. As Polanyi became more aware of meaning being placed outside ourselves, and of the bodily involvement in interpreting clues that reveal an object outside ourselves, he chose to use the anatomical terms "proximal" and "distal." In general, the language is interchangeable between subsidiary or proximal and focal or distal. I prefer the more common terms "subsidiary" and "focal" except where the other language is more appropriate to the thought at hand.

The first type of awareness, the subsidiary, is what we have knowledge of but are unable to tell. We know this only by relying on our awareness of it for attending to the second type of awareness, the focal. We rely on our eyes subsidiarily, to which we pay no attention, in order to see the person focally who we are looking for in the crowd. The painter relies on his/her aesthetic sense to evaluate the correctness of the color or the completeness of the painting. Thus, we use our body, without being explicitly aware of it, to attend to things beyond ourselves.

Polanyi emphasizes throughout his writings that our subsidiary awareness is not the same thing as unconscious, pre-conscious, subliminal, or fringe awareness. He specifies the condition in a footnote in *The Tacit Dimension*, "What makes an awareness subsidiary is the *function it fulfills*; it can have any degree of consciousness, so long as it functions as a clue to the object of our focal attention."[8]

Finally, while the tacit component is made up of that in which we dwell, the explicit is a reality that exists regardless our involvement in it. The two terms of tacit knowing, subsidiary and focal, are both components of the tacit, while the explicit may or may not be held tacitly. Thus, the terms "tacit" and "explicit" are not interchangeable with subsidiary and focal. Polanyi recognizes subjective and objective realities but also discerns that the dichotomy between the two does not exist because we can never know the explicit without the tacit component.

In *The Study of Man*, Polanyi identifies the explicit as our common understanding of knowledge; we can locate it in books, maps, mathematical formulae. The other kind, the tacit, is unformulated. It is the knowledge about which we are unable to tell. If we call the first kind explicit knowledge, and the second, tacit knowledge, "we may say that we always know tacitly that we are holding our explicit knowledge to be true."[9]

The structure of tacit knowing has four aspects: (1) the functional, (2) the phenomenal, (3) the semantic, and (4) the ontological.

1. The functional aspect is the underlying relationship which combines the two kinds of awareness or what I prefer to call the two terms of knowing. It is our attending from the particulars of an entity to the whole they constitute. We attend from the particular features of a face to the whole face. Thus, we may be unable to specify the facial features while recognizing the face. The painter attends from color theory, elements of design, line quality, perspective, texture, light and dark, subject matter, and so forth to the painting. Tacit knowing functions to guide us from subsidiarily held particulars to their integration in a focally known, coherent whole.

This achievement of a coherent understanding of reality is integration not deduction. We cannot formally explain how we know the face, or how we know the painting as "right." The achievement is irreversible; we cannot retrace our steps. To try to do so is to destroy the comprehension attained. We have instead meaningless fragments.

A description of the process of making the quilt *World* (Plate 1) reveals the functional aspect of tacit knowing. I cannot describe exactly from what the quilted image came. I am unable to fully explain why I used a chessboard design integrated with a black and white world map and a circle of gray dolls breaking in and out of the quilt. The chessboard idea is, in part, from my youthful interest in Op Art, my attempts to play chess, and my frustration that life is not black and white. Maybe the train I was riding while I was sketching for the quilt passed a chessboard motif on a billboard. The world map responds to my living in earthquake-ridden California and my interest in plate tectonics; all of the earth's plates are interlinked and no plate moves without affecting the whole. Many have suggested that the people look like a group of skydivers. Yes, I have been skydiving and I may have represented that experience.

When I try to retrace the steps of creating *World*, I am intrigued yet dissatisfied. I am intrigued to find my life sources embodied. I am dissatisfied because the work of art transcends these moments from my life, making this kind of interpretation a disservice to the perception created. What is verbally submitted are a number of disconnected remnants that are remembered and unified in the artwork. While these fragments of my world informed the quilt, they did so tacitly not explicitly. The quilt, *World*, is an integration that originates from but is also more than my encounter with the world. Incompatible particulars are subsidiarily integrated in a new way so as to create focal meaning. Analysis of these particulars disintegrates the integration of incompatibles that is achieved in *World*.

2. In the phenomenal aspect of tacit knowing we are aware of that from which we are attending to another thing, in the appearance of that thing. We are aware of the features of the face in terms of the particular person to whom we are attending. While the painter may not be able to tell how much yellow was used to achieve a certain green, the artist is aware of the strength of the yellow in terms of the green and other colors in the painting.

The phenomenal refers to the shape or appearance the particulars assume upon being tacitly integrated. This integration brings about an awareness of a quality which is not apparent in the dismembered particulars. In the skillful performance of riding a bicycle, the rider is aware of her/his motor activity in terms of the operation of bicycling. Yet the performance of bicycling is not apparent through an analysis of muscle dexterity, centrifugal force, and gravitational pull. If we say that the face has a large nose, we are aware of the largeness of the nose in relation to the whole of the face. Dismembered, the size of the nose is meaningless. The nose is large compared to what? To the face. We are aware of the nose in terms of the whole face; thus, its size takes on significance. We have noted above how the disintegration of the particulars of the quilt *World* destroys the integration achieved and its harmonious quality.

The phenomenal aspect of tacit knowing provides an explanation of our ability to interpret the same set of particulars in different ways.[10] These are evinced in the phenomenon of artistic style, both group and individual.

When a group of artists work from the same model, each artist creates a unique artwork based on that model. As the model points toward an as yet-unknown artwork, and that model is integrated by each artist in the process of creation, it is interpreted differently by each artist. Granted, each artist has a unique physical viewpoint from which to attend to the model, and therefore, the perspective of each artist's work will be, if only slightly, unique.

Bodily involvement is highly operative. An artist two inches taller will encounter the model differently than a shorter person. In teaching drawing, the teacher must physically take the student's viewpoint in order to see what the student sees. A shadow that may be visible from a close proximity to the student will disappear when the teacher takes the student's physical position. The slightest body adjustment can reveal or hide the presence of a finger, flower, or any other given element within a model. Body adjustment or position changes the relationship of the elements to one another within the comprehensive whole of the model. Thus, the artist's bodily encounter with the model is pivotal to his/her experience of the model.

Yet more than physical location accounts for the different interpretation of the same model. The tacit location from which the artist attends to the model accounts for alternative ways of presenting the object of depiction. The artist's unique epoch, race, geography, social history, ideology, training, and so forth affect her/his view of the given model. An artist's tacit location integrated with the model into a new comprehensive whole accounts for artistic style.

Artistic style is individual, as each person's tacit location is unique; thus, a group of artists that may be of the same epoch, race, geography, and training, who are working simultaneously with the same model, will create unique depictions of that model. Yet, underlying these artworks will be a commonality due to their similar, communal locations. This commonality accounts for a cohesive group style such as Gothic, Impressionism, Abstract Expressionism.

Artistic talent is constituted by tacit dimensions of location and accounts for the unique interpretation of a given model. In other words, a person's location becomes an embodied particular in the artwork through practiced performance. An artist's exceptional imagination is guided by skilled attentiveness and insightfulness. These skills are achieved through perseverance and trust in training, practice, and creative effort. An artwork will be more original through dedicated and loyal labor. The particulars will be integrated more creatively, and the discovery of an artistic creation will be more profound.

But an original insight results from more than dedication, loyal labor, and the uniqueness of a person's location. The interpretive framework of a more skilled and appreciative mind for excellence is more perceptive of novelty and its expression. Artistic genius is able to grasp and incarnate the hidden possibilities of integration that the model points toward and that lie beyond the anticipatory powers of current conceptions.[11] The model is a spark to the imagination, a clue to the unknown. Genius moves beyond the expression of current conceptions to a conception of realities yet unknown.

In the creation of an artwork, the tacit integration of the model with the artist's location and skill are, in part, explanatory of the interpretation of the said model in different ways.

The creation of art does not require a model. I persist with the image of a model to maintain cohesiveness in this discussion. However, the artist may approach a painting without an explicit model. Since the model merely points toward an unknown reality, its value to the task is similar to other tacit elements that serve as particulars from which the artist attends to the integration achieved in the artwork.

Like these tacit elements that are largely unspecifiable, a further component comes into play: the original conception of the complex parts that guides the artist toward a comprehensive whole. It is the insight or idea that Susanne K. Langer calls the artwork's "commanding form."

Until the conception of the commanding form, the artist gropes in a vague state, seeking the problem that will lead to the solution or artwork. The commanding form is a fecund matrix that challenges the artist's imagination, for here the problem arises and the artist is challenged to discover its solution. According to Langer, the commanding form is a

> *Gestalt* [that] serves as a measure of right and wrong, too much and too little, strong and weak....It requires such things as ornamentation or intensification or greater simplicity; it may rule out some favorite device of its creator, and force him [*sic*] to find a new one; like a living organism it maintains its identity, and in the face of influences that should mold it into something functionally different, it seems to preserve its original purposes and become distorted from its true lines rather than simply replaced by something else.[12]

While the artist works with the artwork, the commanding form guarantees correct judgment. It is that non-specifiable element that guides the artist in the struggle to find the exact form of expression needed to incarnate her/his original conception. The artist knows her/his aim only in broad terms and must rely on a sense of the commanding form, a deepening and innovative coherence, to guide her/him toward the artwork's embodiment.

The commanding form beckons the artist to a new vision of reality which in turn presents the artist with the problem of its expression and the discovery of its form. The artist may discover the commanding form before lifting a brush to the canvas; this occasion causes the artist to state that s/he was grasped, badgered, possessed by an idea that was relentless until expressed. Yet the commanding form is embryonic and does not attain final form until the process of creation has been completed.

The idea may be developed in many possible ways, and the artist can best be understood as a seeker led by an indeterminate unknown. The artist may gropingly begin a painting and through the process of painting discover, with an intuitive "click," the commanding form. This discovery may also be the moment when the artist recognizes the painting as complete.

Within the phenomenal aspect of tacit knowing, the particulars which are tacitly integrated are relied upon to focus on a quality, or more precisely, a commanding form not apparent in the particulars themselves. Thus, the ability to perceive certain phenomena as not-yet-known requires preparation, preoccupation, and an openness on the part of the knower.

This is evident in Weismann's pursuit of discovery. Weismann took great measures to prepare his laboratory, a remote and simple house in the Mexican countryside, and to situate himself within it. Then he opened himself to dwelling in the mass of experience that was his life in hopes of discovering a story.

When the artist creates, an heightened awareness occurs due to a long period of training and the personal effort. Particulars from the artist's location are incorporated into her/his artistic repertoire which enables her/him to rely on them and to grasp a quality of their integrated appearance in a manner not appreciated by someone whose awareness was not so trained. Or when someone practices artistic creation without the benefits of formal training, s/he will be able to express a quality in her/his painting unavailable to someone who has not cultivated an openness to seeing and making art. Often such art is classified as "folk art." The folk artist has not mastered the skills only training can provide, but does achieve works that are original in their conception.

Since the phenomenal aspect is an outcome of the functional structure of tacit knowing, it follows that if the commanding form of the artwork is not respected, the organic unity of the painting will be lost and the piece will fail. While you attend to the particulars focally, the particulars cannot function subsidiarily, causing the integration of the particulars to be destroyed. More than likely, if you attend to your muscle dexterity while riding a bicycle, you will fall off the bicycle. Or when looking at a face, if the nose is focused upon, it will lose its quality as a large nose. If the artist attends to the model, rather than the unknown it points toward, the creative integration that points toward the anticipated painting will be destroyed.

3. We may know a face without being able to specify its particulars (functional), yet we are also aware of the features of that face in terms of the person to whom we are attending (phenomenal), and at the same time we rely on our awareness of the face's features for attending to their joint meaning or mood. This reliance on tacit elements to focus on meaning is the semantic aspect of tacit knowing.

Characteristic of the semantic aspect is that meaning is displaced away from ourselves, thus the use of the two terms of tacit knowing: proximal and distal.[13] We attend from our tacit knowledge of the proximal term to our focal knowledge of the distal term. It is through the integration of particulars which we have embodied that the reality these particulars comprise is endowed with meaning. This meaning is grounded in our reliance on the framework or location in which we dwell.

We know explicit reality through our bodies. William H. Poteat has coined the term "mindbody" to be descriptive of our tacit reliance upon our bodies, minds, spirits, and locations which are integrated and serve as the bedrock of all our knowing and being. When we speak, we gesture. It is all but impossible to express our thoughts apart from bodily movement. Likewise, a gesture in itself is meaningless, but points toward our mental activity. We come to know a person's mind through that person's bodily gesture. Ultimately, our mindbodily being in the world "is opaque to reflection..., [and it is not] something given once-for-all, static, fixated."[14] We live mindbodily in the world, open to change and discovery. To attempt to separate and focus upon specifics of our minds or bodies is to dismember their integration and destroy their meaning. Poteat's term "mindbody" is rooted in Polanyi's theory of tacit knowing. All knowledge is grounded in mindbodily indwelling to make sense of the outside world.

Indwelling is our extending ourselves into something known so that it becomes an extension of the self and a tool for further exploration of reality. When a blind person uses a white cane for the first time, s/he will be aware of its impact against her/his fingers and palm. But as s/he learns to use the cane, the awareness of its impact on the hand is transformed into a sense of its point touching the object being explored. S/he becomes aware of the feelings in the hand in terms of their meaning located at the tip of the cane. What was initially a meaningless feeling, the impact of the cane upon the fingers and hand, is transposed into meaningful data after practice and integrating the cane into the mindbody. The meaning is displaced from the hand to the tip of the cane.[15]

When a painter first picks up a paint brush s/he focuses on the brush as an object, but once the painter acquires the skill of using the brush, the brush becomes an extension of the painter's mindbody so that the painter may attend from the brush to the painting. The painter dwells in the brush, reaching out from this known entity in order to discover an unknown, a new reality, the painting. Weismann and I dwell in our life experience, what we know, in hopes of discovering, creating, something new.

The particulars, like the brush, in which we dwell are clues for extending ourselves into the discovery of external reality; by doing so we find meaning not in the particulars, but in their integration. The merged set of particulars point toward a reality which is away from ourselves. Human knowledge cannot be justified by explicit procedures, nor does it rest on caprice since it is dependent on the particulars it embodies. Meaning arises through the integration of the particulars. "Nothing said, written or printed, can ever mean anything in itself: for it is only a *person* who utters something—or who listens to it or reads it—who can mean something by it."[16]

The semantic aspect brings the issue of location to the fore. It is through being a particular person within a particular culture that we are limited, but only through dwelling in our locations are we able to transcend our limitations. Our historical context is our tool for discovery. This is why art is culturally bound. The artist dwells in his/her world and integrates its particulars in a new way to embody meaning. Polanyi has given us a dance between the tacit and explicit in which knowledge and meaning emerge within the relationship of the two. He makes true the statement, "it takes two to tango!"

A person commits to dance with the unknown. From the dance emerges the known. The known is a whole integrated by the person from all that the person is. The known is a part of the unknown. Knowing is triadic in structure: (1) the focally known or explicit, (2) the person's world held subsidiarily, and (3) the person who integrates a meaningful relation of the subsidiaries to a focal whole.[17]

Even a solitary act, as the creation of art often is, is triadic and, therefore, fundamentally social. The artist as knower is always a part of a socio-historical context as described in the preceding section on World. As the artist brings his/her location subsidiarily to the artwork, the community is involved in the creation process. The artist never acts alone but always from the community in which s/he dwells and to that community for its acceptance.

4. The ontological aspect of tacit knowing tells us of what tacit knowing is a knowledge. Since tacit knowing establishes a meaningful relationship between the two terms, the ontological aspect is our

understanding of this "from/to" relationship as a comprehension of what is real. We tacitly integrate into our mindbodily existence particulars of reality that lead us beyond ourselves to an aspect of reality's comprehension. The ontological aspect is the embodiment of a meaningful comprehensive entity. All of our knowledge unfolds within a context tacitly held by a person—a context that includes our being in the world mindbodily.[18] When we make a thing function subsidiarily, we incorporate it into our mindbodies so that we come to dwell in it. We do not passively observe things, but actively dwell in them in order to understand reality.

Polanyi explains how our bodies are the ultimate instrument of all our external knowledge: that the way we come to know reality is determined by our mindbodily efforts. We attend from internal bodily processes to the qualities of things outside. As we do so the meaning of those qualities specify our internal processes. Knowing is transposed away from ourselves to external reality.[19] We rely on our awareness of our mindbodily contacts with things outside our bodies for attending to them. Our mindbodies are the means through which we attend to and come to know our world.

We know our bodies not as objects but as the instrument through which we know. When we use some things, such as the blind person uses a cane to find her/his way, or a painter uses a paint brush to attend to a painting, the thing, in this case a cane or paint brush, changes its appearance to become like our bodies. We know, we attend to reality through that thing. We incorporate particular entities so that as subsidiaries they extend our bodily knowing, or more accurately, our mindbodily knowing.

Thus, the act of tacit knowing results in our comprehension of something real which may manifest itself inexhaustibly in the future. Polanyi defines reality as that which "may yet reveal its truth in an inexhaustible range of unknown and perhaps still unthinkable ways."[20] Implied in this definition is that our knowledge is in process. An indeterminate range of perceptions bearing on reality are present.

Furthermore, the tacit dimension of knowing may contain actual knowledge that is indeterminate. Subsidiary awareness, integrated through our mindbodies, and actively employed in the act of discovery and knowing reality, may not be explicitly identifiable. We know, not because we can articulate something, be it a maxim, formula, or some other type of description, but because we rely on and employ that knowledge. Knowing is not reflective but active.

Tacit knowing depends on our focus of attention and in this sense includes caring. A face is sought out in a crowd because we care about the person sought; the artist cares about expressing her/his vision of reality. Underlying knowing, then, is a desire to know and a belief that there is something to know. When posed with a problem, the process of discovery is engaged. The knower dwells in a reservoir of subsidiaries made up of socio-historical context, intellectual frameworks, aspects of reality which point beyond themselves to an integrated whole. When the pool of tacit knowing is impacted by something new, novel understandings are perceived. In this way, reality may reveal itself indeterminately in the future.

Motivated by desire and caring, the artist dwells in the framework of the arts and through an active mindbodily engagement with other artworks (her/his own and/or other artists' and art forms) that bear on reality, the artist becomes inspired. A potential yet-unknown artwork is sensed. Creativity begets creativity.

The artist grasps an intimation of a coherence of particulars not yet comprehended. S/he has a tacit foreknowledge of an as yet-undiscovered artwork. Weismann wanted to create a story and had a hunch that it would be discovered if he would but follow the desire and the clues upon which he was brooding: placing himself in an art-making laboratory, thinking of a former friend, looking out a frosty window. In creating *World*, I was pondering the anti-apartheid demonstrations, the rules of quilting, the complexity of Christian ethics. These thoughts united with experiential subsidiaries became positive aids to artistic perception. The merging of otherwise disparate elements transmitted clues to a work of art yet to be made.

Unknown reality attracts the artist's attention through clues that invite, even demand, inquiry. Weismann was rewarded by a paragraph that began as a letter to a friend and turned out to be the beginning of his story. I began making a black and white quilt that became a three-dimensional artifact. The paragraph and *World* were unexpected, but their truth is recognized by the artist when the form emerges. Through encounter with his/her world, the artist is beguiled, even harassed, to make form, to embody a reality that exists outside of ourselves. The commanding form attracts us to its incarnate discovery. The work of art becomes real, is appreciated for its truth, because it draws us on, even obliges us to search and discover, and it compensates our efforts by revealing more, unexpected, yet recognizable meaning.[21]

In the creative process the artist sets out intuitively with an aim which is largely indeterminate. Weismann aimed to write a story but was unsure what the story was about; I intended to make a quilt but did not perceive its final form. As the imagination is let loose to ferret out a path of possible clues, the artwork unfolds, takes shape, and the artist's aim is altered by intuitive feelings to better fit the means that is providing the solution. The artwork is said to take on a life of its own. Finally, the idea is fulfilled. The commanding form is realized while it tacitly guides the artist to its expression through form.

The artist's quest is thus guided by a glimpse of a tacitly foreknown integration. Clues originally suggested by her/his caring, desire, and commitment to the current art scene, point to a real presence. The commanding form sustains the vision to the artwork's completion. If the creative hunch proves to be a good one, the artist is able to incarnate a new coherence which s/he expresses with universal intent.

"Universal intent" is a term Polanyi uses in *Personal Knowledge* to refer to our conviction that contact with reality has been achieved through serious and impassioned acts of knowing. This conviction might not prove to be correct; Polanyi frequently reminds the reader that knowers are fallible. To advance knowledge claims with "universal intent" indicates that the knower believes that his/her vision is consistent with (uncovers) reality and that this vision is available and illuminating for other persons. "Universal intent" indicates that a person's knowledge is open to correction and enlargement but is backed by the belief that others can and should recognize the realities it points to.

In the case of art, others participate in the event the artwork embodies and recognize it as true. Yet the artifact, being real, may reveal itself to the viewer's eyes in an indefinite range of unexpected manifestations.

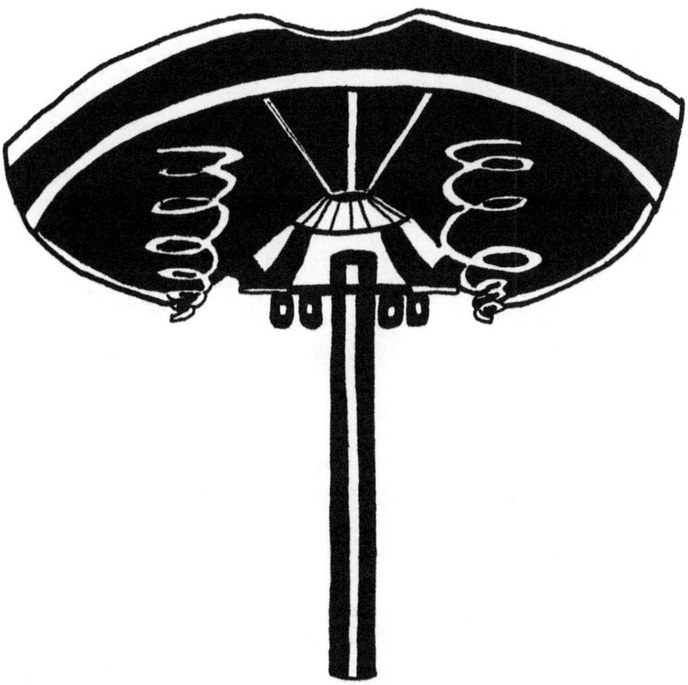

NOTES

1. Donald L. Weismann, "In Pursuit of Discovery" in Thomas A. Langford and William H. Poteat, eds., *Intellect and Hope: Essays in the Thought of Michael Polanyi* (Durham, N.C.: Duke University Press, 1968), pp. 387-388.

2. Rollo May, *The Courage to Create* (New York: W. W. Norton and Company, 1975), p. 46.

3. See especially Michael Polanyi, (1) "The Abstract Arts," (2) "Dwelling In and Breaking Out" in *Personal Knowledge* (*PK*) (Chicago: University of Chicago Press, 1958; corrected ed., 1962), pp. 193-202 and (3) "What Is a Painting?" *American Scholar* 39 (Autumn, 1970), pp. 665-669.

4. Michael Polanyi, *The Tacit Dimension* (*TD*) (New York: Doubleday, 1966; repr., Gloucester, Mass.: Peter Smith, 1983), p. 24.

5. Charles S. McCoy, *When gods Change* (Nashville: Abingdon, 1980), p. 92.

6. Weismann, "In Pursuit of Discovery," p. 390.

7. Michael Polanyi, *Science, Faith, and Society* (*SFS*) (repr. London ed., 1946; Chicago: University of Chicago Press, 1964), p. 15.

8. Polanyi, *TD,* pp. 95-6.

9. Michael Polanyi, *The Study of Man* (*SM*) (North Staffordshire: University College, 1959; reprint Chicago: University of Chicago Press, 1963), p. 12.

10. See John Apczynski, *Doers of the Word: Toward a Foundational Theology Based on the Thought of Michael Polanyi* (Missoula, Montana: Scholars Press for the American Academy of Religion, 1977), pp. 104-5. I have relied on his outline of the phenomenal aspect of tacit knowing throughout this section.

11. See Michael Polanyi, "Problem Solving," *The British Journal for the Philosophy of Science* 8 (August 1957), p. 94.

12. Susanne K. Langer, *Feeling and Form* (New York: Charles Scribner's Sons, 1953), p. 122.

13. Polanyi, *TD*, p. 13.

14. William H. Poteat, *Polanyian Meditations: In Search of a Post-Critical Logic* (Durham, N.C.: Duke University Press, 1985), p. 18.

15. Polanyi, *TD*, pp. 12-13.

16. Michael Polanyi, *SM*, p. 22.

17. Michael Polanyi, *Knowing and Being* (*KB*), Marjorie Grene, ed. (Chicago: University of Chicago Press; Phoenix Books, 1969), pp. 181-183.

18. Apczynski, *Doers of the Word*, p. 107.

19. Polanyi, *TD*, pp. 13-14.

20. Polanyi, *KB*, p. 141.

21. Lady Drusilla Scott, *Everyman Revived: The Common Sense of Michael Polanyi* (Lewes, Sussex: The Book Guild Ltd., 1985), p. 68.

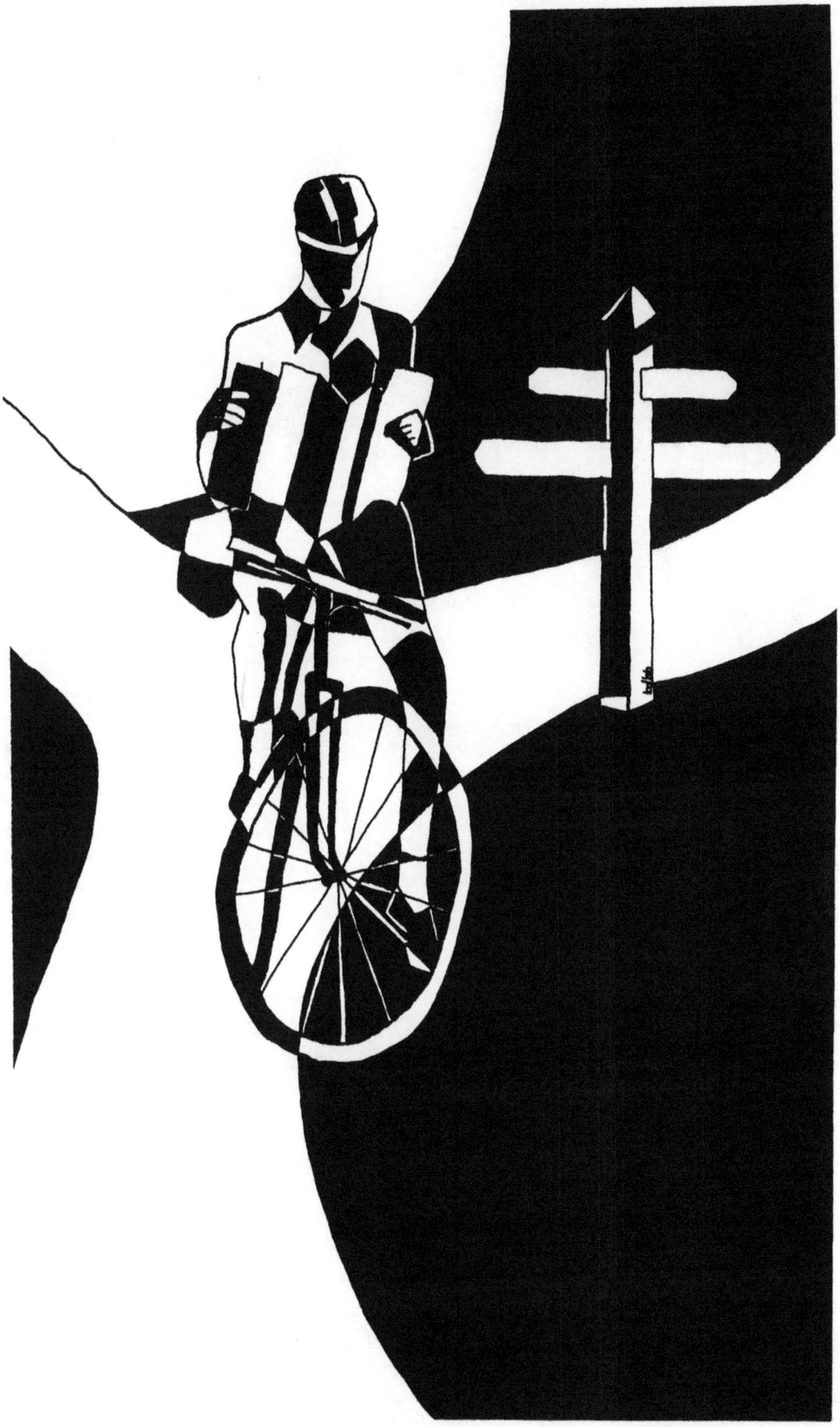

Chapter Two

Committed Indwelling

1. Commitment

For Polanyi, commitment is our assertion of what we know and do with universal intent. Polanyi's thesis of commitment is the personal and responsible compliance of the mindbody to the requirements of a reality independent of it. A person entrusts himself/herself to reality and proclaims knowledge in response to making contact with that reality.

Polanyi overcomes the subject/object split. Commitment is not a subjective endeavor, as the mindbody is submitted to an independent reality; yet, because this submission is guided by individual passions, it is not strictly objective. Within commitment, the personal and universal require each other as a commitment from ourselves to reality. We can never know reality completely; thus, such a commitment is always risky: it is a venturing out from our world to find the unknown.

Polanyi cites Christopher Columbus as an example of someone who had the courage to be committed, risking the unknown and the possibility of being wrong as a necessary element in discovery. Columbus staked his life and reputation on a belief we now recognize as a small distorted fragment of the truth, but it impelled him to move in the right direction. His discovery was not that originally intended, to reach the East by sailing West, but a discovery of a new continent.

Or, for example in the case of Weismann: he committed himself to the creative process by withdrawing to a selected place with only the tools of his trade for occupation. He risked a year of his life to perhaps return home empty-handed. He was beguiled by the urge to create—he had a hunch the creation would be based on his life experience. He knew no more. He set sail across the waters of his experience to discover an as yet-unknown story that he could recognize as true.

The experiences of Columbus and Weismann are examples of the risk and daring involved in discovery. When we commit ourselves to a glimpse of reality, it may be mistaken, and it is certainly incomplete.

What we discover may not be what we expect but may open up aspects of reality we never imagined. We may bring something new into our midst.

Commitment, while being objectively oriented toward universal intent, is always personal. Columbus could not blame his maps, the rules of the sea, or any other structure for his failure to reach the East; yes, these were tools, but he personally was responsible for their integration. Columbus's act of judgment led him in his course. Commitment does not bind the discoverer but frees him/her from maxims or structures. It allows for an outbreaking into the unknown, new, unexpected.

Discovery claims to make contact with reality beyond the clues on which it relies. Personal knowledge commits us, passionately and far beyond our comprehension, to a vision of reality. Formalizations cannot relieve us of commitment. "For we live in it as in the garment of our own skin. Like love, to which it is akin, this commitment is a 'shirt of flame,' blazing with passion and, also like love, consumed by a devotion to a universal demand."[1] On the whole, the work of the knower (scientist, explorer, artist, even worshiper) is so workaday, humdrum, and sedentary that only the deepest motivations can inspire the kind of discipline it takes to meet knowledge's demands. Commitment is a discipline born of love.

In my work as a painter, I have found that as I remain attached to structures of painting or ideology my paintings fail. I do not create the new, but usually a poor rendition of what has gone before. I achieve illustration at best. What has happened is that I have been unwilling to risk commitment; I have taken the safe route of relying on structure. At times this may prove to be color theory, or design, or Christian ideology.

Often when I, in the name of love (and this is a common case for many contemporary Christian artists), try to maintain an allegiance to a specific scripture or image of Christ, I find just the opposite occurs. I fail to capture the essence of what I am attempting to portray. I have refused to commit, to risk engaging with the problem at hand. What I have mistaken for commitment has been an avoidance of personal responsibility.

When I let go of the structure, when I am courageous enough to engage personally with the object of intent (painting), my paintings often succeed. The amazing thing is that as I have let go of a conviction in order to paint, my painting unexpectedly presents an insight into just what I have formerly been stifled by.

An understanding of how or why this is so is evident in Polanyi's epistemology of tacit knowing. When I attend to a conviction I am not then free to attend to my painting. Painting, in this case, is a subsidiary act from which I attend to my conviction, and so I produce illustration. When I risk commitment to a painting, I attend toward the unknown from that in which I dwell: my convictions, paint theory, brushes, etc. A conviction, theory, or tool cannot be used by attending to it but is employed by attending from it to what we are trying to achieve. To use something means having to commit *from* it *to* a reality beyond oneself. This commitment from who we are and all we dwell in to a reality beyond is the difference between personal and subjective.

Commitment is a term that Polanyi emphasizes in *Personal Knowledge*, but in the Introduction to *The Tacit Dimension* (p. x) he suggests that he has worked out more carefully the structure of tacit knowing; talking about overt commitment is no longer necessary. I find, however, that overt commitment discriminates between private hobby art and art therapy on one hand, and public "professional" art on the other.

Commitment requires us to practice the *via negativa*—or the opening of ourselves to the unknown. By practicing the *via negativa* we can come into relationship with a reality of which we are previously unaware.

The *via negativa* has been developed in Eastern Orthodox Christian theology and involves the setting aside or loosening our thought forms from those previously held so that revelation, the new, may occur.[2] It is to shift focal awareness to subsidiary awareness so that we may attend to the unknown: a heuristic leap. In order to paint, to create new form, we must set aside all that binds our mindbody to previously known forms so that new forms may emerge.

Commitment is a loosening of our fetters, a movement into freedom. It is allowing the ego to step aside, submitting to reality and accepting what comes and the responsibility that comes with it. This often scary step blocks many people from creativity. To remain uncommitted is to remain sure or safe. A story-image will be helpful here.

It is night. A woman stands in a familiar well-lit room. The light illuminates the comfortable room including an adjoining door. The door is slightly ajar. The room beyond the door is couched in darkness. Moonshadows stream through a window and dance across the room. The woman is uncomfortable peering into the dark room. It is a room she has never entered but it arouses her curiosity.

The light switch is across the large dark space on the opposite wall. She wants to enter the room but an uneasiness creeps in. What might she encounter as she progresses into the darkness intending toward the light? The moonbeams beckon her to come dance with them. They will guide her way to the light switch. If she enters the darkness, she will no longer be sure of her whereabouts. She will have to trust the tentative light the moon offers and the light from the familiar room that streams in when the door is pushed open. She will not be able to see clearly the walls or the room's contents.

She is faced with risk—if she chooses to move into the darkness, she may encounter injury or be stopped by the furniture or a wall. The moonshadows suggest hints of form. They offer dim clues of the room's contents. Her imagination and the moonshadows may guide her, play visual tricks, or even stop her progression across the room. A chair may offer a clue to movement, intimate an illusion, or cause a fall.

In the dark, dancing with the moonshadows, she is free to imagine. In her present well-lit location she is safe and sure. The steps into the dark, unfamiliar room are a movement into mystery, the unknown. It is a responsible act, to choose to enter the unknown in order to encounter it, and to move cautiously so as to not be injured or carried away by false imaginings until the light is reached.

The imagination is invoked through entering the darkness and guides her (our) path to reality. Polanyi explains the workings of the imagination as a perceptual guide.

> Perception works within our imagination, and when there is a difficulty in making out what it is that we see, we send out thrusts of our imagination to explore what that thing may be. And, of course, as it is capable of finding truth, the imagination is capable also of error: it can produce illusions.[3]

The woman may discern furniture in the dark room (reality) or perhaps a phantom shape (illusion). Both the seeing and object of perception function by stages as we endeavor to see. When the woman's eyes adjust to the darkness and she is more easily able to detect form, she moves through degrees of perception or illusion. In the darkness, the imagination works toward discernment, recognizing what is really before us, or illusion, seeing things that are not there.

Creative commitment requires that we enter the darkness, silence, mystery, trusting in the shadows of light that reality offers. We yield to the creative process, believing it will reveal a perception of reality, not illusion. Commitment requires we let go of all that binds and enter the void so that creation may emerge.

We do not create but submit to creativity by setting aside control. In the creative act, the artist is not in control but sets aside control in order to dance with the unknown. It is in the dance that the new forms emerge. A painter cannot will a painting but only risk one. Weismann could not will his story but only risk one. Likewise, a mystic cannot will a revelation but only risk one.

2. Dwelling In and Breaking Out

Since our original definition of creativity is the process of bringing something new into being through the active and committed encounter of the individual with her/his world, creativity may be understood as a dwelling in and a breaking out. We can say that the process of creativity includes a journey, a promise of dimly sensed form, and a battle.

Myth opens a view of the universe and makes us feel at home in it.

It gives us access to a detached experience of meaning, echoes of which will follow us into everyday life. Myth makes us aware of a deeper reality we deem to be "true" by evoking in us an experience we hold to be genuine.[4]

Two myths in the Bible provide maps for our pursuit into creativity. First is the Exodus account in which Moses leads his people out of bondage toward the promised land. The Israelites wander forty years in the wilderness before arriving at the promised land. The promised land was largely unspecified but the people knew it when they arrived. To attain the land, the people had to engage in battle. In the synoptic Gospels of the New Testament, Jesus journeys into the wilderness for forty days and nights. He responds to a promise of messiahship yet does not know exactly how that promise will manifest itself. He does battle with Satan and through the battle recognizes the essence of his call.

The artist gropingly but effectively journeys toward the promise of a dimly sensed form through indwelling. Indwelling is the interiorization of subsidiary particulars in the act of tacit integration. Interiorizing particulars means we "assimilate" them into our mindbodies. So indwelling is the personal participation of the knower in what s/he knows mindbodily. Indwelling underlies all observations because every particular that functions subsidiarily is dwelt in. Thus, tacit knowledge is held by indwelling: the parts of a whole are comprehended subsidiarily in terms of that whole. In other words, we grasp meaning by dwelling in particulars. The artist makes a new creation through joining elements that point toward the expressive form and embody meaning through dwelling in those components mindbodily.

To interiorize is to identify ourselves, our bodies, with the whole in question by making its particulars function subsidiarily. We rely on our mindbodies to attend to explicit reality so that all knowledge is mindbodily rooted. We dwell in all that we have learned, interiorized, or assimilated mindbodily over the course of our lives. We dwell in our worlds.

The painter relies on color theory for understanding the process of painting. S/he attends from the theory mindbodily to the painting and is aware of the theory while using it in terms of the painting it serves to explain. The color theory is learned through its practical application.

Polanyi points out that true theoretical knowledge lies in our ability to use it.[5] To be a wine connoisseur, for example, is futile and meaningless if one does not taste wine; indeed, the connoisseurship of wine tasting cannot be achieved outside of actually tasting many wines. Indwelling is being-in-the-world mindbodily; it is the fundamental mode of being human that tries to understand.

The theory of indwelling recognizes that the integration of the parts foundational to tacit knowing is participatory. We actively engage in what we know through interiorization. We understand reality not by looking at it but by dwelling in it.[6]

Indwelling is critical to the task of knowing, and the process of assimilation includes an alternation between analysis and integration. Let us consider the making of art. The artist, we have learned, does not focus on the particulars that point to the as yet-unknown painting, but dwells in them mindbodily in order to achieve meaning. The use of an object or model as a pointer toward form can be helpful in the process of expression, but it can also be binding.

A case in point is my pursuit in learning to draw human form. When I first began life drawing, I drew abstract forms that were intense and emotionally charged. My teacher at the time told me to "never learn to draw" lest I lose my fresh style. As I continued in art school, another teacher insisted that I learn how to draw the figure accurately. When I rebuffed him by quoting my former teacher, he replied, "abstraction, to be successful, must be achieved out of knowledge not ignorance." I acquiesced and allowed him to teach me the technique of life drawing. He was an excellent teacher, and I learned easily. In a matter of weeks I could draw anatomically correct proportional figures. By focusing on the particulars of the figure in order to draw it, the original meaning of my figures was destroyed.

Ever since, I have been trying to recover abstraction. I will never be able to return to my original style. My new style should ideally be deeper and more powerful than the earlier one. However, the process of learning to draw has so obscured my original indwelling in the figure, that I have been unable to achieve a meaningful abstract style. Yet, through practice and perseverance, I believe I will come to a deepened style.

It may be better to abstract out of knowledge, but it is also more difficult. Knowledge, usually an asset, can impede breakthroughs if we focus upon it rather than dwell in it. The explorer leaves home for the journey of knowledge. The knowledge gained, a battle ensues which oscillates between a focus upon an exacting system and self-surrender to discovery before the promised land is gained.

The process of focusing on a comprehensive entity, next shifting our attention back to focus on the particulars, temporarily destroying the comprehensive entity's meaning, then to again focus our attention away from the particulars on to their joint meaning, is a foundational aspect of the creative journey. This alternation between analysis and integration leads progressively to a deeper understanding of the focal whole. The loss of meaning that results from analysis of the particulars is usually of less consequence than the understanding gained by their temporary dismemberment and identification. The successive new integration of the particulars carries subsidiarily the deeper understanding gained through analysis. Even though the subsidiaries' particular identity is effaced in the comprehensive entity's integration, our understanding of the particulars is enriched by their participation in the whole.

The movement between analysis and integration is managed by the knower who can keep the alternating focus in mind while attending to the other. This is the way knowledge advances and discovery occurs. The dimly sensed promised land draws the seeker on; the seeker pauses on the journey to check maps and other specifics of travel in the light of the distance travelled. Once again the seeker continues on the journey. This process of travel and pausing to check the maps leads the seeker to the promised land. Once the promised land is reached, the seeker dwells there. S/he becomes a new resident unable to return home as the same person. The land of origin takes on a new, deepened meaning in light of the attained promised land.

Human beings, as inherently active knowers, are unique individuals making themselves since the beginning of life. The whole self includes the biological and social being, tradition, and all that we dwell in and extend ourselves into. The whole, active, knowing self with its unique store of tacit knowing is involved in a process of building up and innovating.

Lady Drusilla Scott succinctly explains that learning is a combination of two opposite tendencies, "the building up of frameworks to which new experience is assimilated, and the adaptation of these frameworks to accommodate new experience."[7] Polanyi calls these tendencies dwelling in and breaking out. Our ability to hold these tendencies together establishes our reality as a person and our relation to truth.

The artist, in order to give truth form or to break out, shifts her/his attention from integration to analysis and back again, while keeping a hold on both. When an artist paints a portrait, s/he attends to the joint meaning of the painting, then back to the particulars of the face. S/he notices the shape of the shadow below the nose, the relationship of the width of the mouth to the eyes, then s/he focally shifts back to the painting as a whole. If the artist should keep her/his attention on the particulars, illustration will be achieved at best, and more than likely a dis-unified composition will result. If the artist attends solely to the integration, his/her grounding will be lost. A piece divorced from experience will result: it will be meaningless and illusionary.

Novel integrations enable the artist to break out from concepts that can be binding, while the concepts offer stability and continuity with experience. This is the battle before journey's end. A dimly perceived form lies ahead, but the artist is bound by existing frameworks. A shift in focus, a submission to the artwork and its commanding form, breaks the framework to reveal something new. Progression toward artistic form includes a struggle between analysis and integration; if either dominates to the exclusion of the other, artistic achievement will have been lost. Form that is "true" results from a wrestling in which the two tendencies find harmony and balance together. In the achievement of mutuality, the dimly sensed form, the promised land, is attained. The journey, at least temporarily, is at end.

The human mind is essentially restless. It calls into question previously achieved satisfactions. We are not satisfied to rest in a problem solved. The artist is not satisfied to stop in a completed artwork. The creation achieved beckons the artist to a new as yet-unknown form. New problems urge us on to be pondered, new paths beckon to their solution. This may be as simple as crossword puzzles or jigsaws, or as complex as scientific discovery or the creation of new form. We have a never-ending urge to break through our fixed frameworks. The artist is driven and responds to a continuous call to break out of discovered form. The journey never ends, the dimly sensed promise is never realized, the battles to break through our frames never cease. The call to create form and discover meaning never concludes.

Polanyi states that the most radical manifestation of our urge to dwell in and break out is the ecstatic vision of contemplation. We dissolve the screen of our conceptual framework. We cease to move through experience and instead pour ourselves into experience; "we cease to handle things and become immersed in them."[8] The experience is vivid yet dream-like; it is timeless and without definite spacial location. As Polanyi explains,

> It is not an objective reality; for it is not the focus of an intelligent perception anticipating future confirmation by tangible things, but resides merely in the colored patches of various shapes which the things present to the eye.[9]

The person participates, indwells completely in what is contemplated. In contemplation the knower resides intuitively in natural reality in order to experience an imageless and nameless reality. Time, space, and discursive thought are transcended into an awareness of a highly integrative coherence bearing the meaning of ordinary experience but in an extraordinary way.

The contemplative dwells in natural reality through the mindbody, but rather than being aware of the objects of nature, s/he is aware of a deeper reality that underlies nature's meaning. Contemplation, like all feats of knowing, is a skill. The contemplative indwells a religious framework. When that framework is intuitively integrated with clues from nature, religious meaning is revealed. In the act of contemplation the natural world reveals extraordinary meaning.

Lady Scott names this phenomenon of nature revealing a deeper meaning "supra natural." It is not to be confounded with the term "supernatural." Supernatural suggests a dualistic world in which the world as we know it is broken into by another world that does not obey the laws of nature. The supernatural is arbitrary; it suggests magic and annuls natural reality. The supra-natural can only be glimpsed through the clues offered by nature. When it is grasped it reveals a more profound understanding of our world. The supra-natural is not separate or anarchic, but is continuous with nature and its laws.

The discovery of the supra-natural is the same process of tacit integration common to all knowing. The natural world is dwelled in and functions as a clue to its meaning. The supra-natural does not render the natural meaningless, but it is a more comprehensive meaningful integration than natural meaning.

However, the supra-natural is less comprehensive than the Supra-Natural meaning that religion reveals. Lady Scott portrays the supra-natural as the meaning of the natural just as the mind is the meaning of the brain.[10] Her analogy may be furthered: even as the mind is the meaning of the brain, the brain is itself a meaningful entity. The mindbody is a deeper coherence of meaning which includes the mind, brain, and all other aspects of who we are as embodied beings in the world. Thus, while nature can be meaningful in and of itself, the supra-natural reveals a deeper coherence of meaning not ordinarily grasped, and the Supra-Natural illuminates the whole of our world.

In an aim to embody the supra-natural, the artist, like the contemplative, ceases to handle things and instead becomes immersed in them. When an artist is questioned about the process of making art, the artist most often states that it is a timeless yet vivid experience. It is for the artist a spiritual moment of vision. Like the contemplative, the artist does not focus on nature but pours herself/himself into a communion with nature. But unlike the contemplative who indwells intuitively, the artist dwells into nature imaginatively and anticipates the tangible. The artist resides in the colored patches of various shapes which are present to the eye, not as a means of communion with nature (though this may occur), but in order to make the supra-natural visible, and thus tangible to the eye.

The artist makes visible imaginative tacit integrations which are wholly non-specifiable. The artist indwells an artistic frame which aims toward a visual embodiment. When this frame is imaginatively integrated with clues from nature, aesthetic form is presented. What art creates, then, is not illusion but incarnations of the artistic imagination.

Polanyi asserts that art is not an illusion because "the arts do not exhibit things that could be really there and yet are not there; they exhibit things of a kind that cannot exist, either in nature or among men [*sic*]."[11] Art is not an illusion because it does not pretend to be something that it is not.

Polanyi's description of what art does—exhibiting something that cannot exist—can be understood in terms of nature. Polanyi understands art to integrate incompatibles. For example, in a painting the flat surface of the canvas is naturally incompatible with its perspectival depth. While perception and illusion integrate elements that belong together naturally, the elements integrated by art are not integrated naturally in nature or in humanity. The achievement of artistic integrations is a heuristic leap that embodies imaginative integrations which incarnate what cannot be communicated otherwise: the supra-natural.

Art unifies elements that communicate and others that contradict communication. The quality of art's harmonious form is not altogether found in ordinary reality, and consequently it does not communicate factual information. Polanyi explains:

> An artistic problem is the imaginative anticipation, not of unknown facts that already do exist, in some sense, in nature, but of a fact of the imagination—of a poem or a painting that *could* exist.[12]

Art draws on our experiences of being-in-the-world mindbodily and embodies them in its configuration.[13] Art does not imitate nature; it creates something new which does not occur in nature so that it can make the supra-natural visible. Art conveys not facts of ordinary experienced reality that can be observed but novel tacit integrations or facts of the imagination that are indwelled. Thus, the artwork created is generated from experienced reality, yet it enlarges our experience of reality by integrating ordinary reality into an highly imaginative coherence. This coherence, art's reality, which relies on yet moves beyond ordinary coherences of reality, I call supra-natural reality.

In naming the reality art presents "supra-natural," I do not propose a Romantic notion of art. Art is not an apprehension of reality superior to scientific or linguistic understandings. Supra-natural reality is not higher than or above ordinary reality. The supra-natural is a continuous yet deeper understanding of ordinary reality. Natural reality is an essential clue to the supra-natural which integrates natural reality in an extraordinary way. The supra-natural is always in relation to the natural, as it can only be grasped by passing through the natural. The term "supra-natural" is not intended to be hierarchical. It signals apprehensions of reality which exist through and beyond normal understandings.

Art's distinctive quality is that it embodies the supra-natural in a tangible way. The visual arts shape ordinary space extraordinarily by rendering it visible and thus tangible. Achieving this extraordinary perception of reality requires a great degree of imaginative effort. The distinction between natural and supra-natural integrations is not sharp; it is one of degree. Art requires a greater degree of indwelling or personal participation than does science or language.

The artist dwells in clues imaginatively to break out toward new form. The process of dwelling in and breaking out is pivotal to all acts of understanding. All knowing is tacit or rooted in tacit knowledge, whether in science, humanities, art, or religion. Our sensing of the scientific endeavor as being different from art, and our sensing of a religious dimension to art are varying degrees of personal participation answered in Polanyi's theory of indwelling. While all knowing is the same, by the same process, always personal, always mindbodily, always social, never definitive, based on our belief in it to be true, we do not dwell in all types of knowing equally.

Science has claimed to be personally detached, to be strictly objective; religion has been accused of being purely subjective without support from objective reality. These are false dichotomies but are believed by many to be true because the scientist maintains a screen of observation and manipulation while the religious abandons it completely. The scientist often dwells in reality to a lesser degree than the religious. Both indwell, both submit themselves to reality—both break out of their frameworks and are changed—the process is the same, only the degree of personal participation differs. This gives us the illusion that one is objective while the other is subjective.

Therefore, according to Polanyi, the degree we participate in, or dwell in, subsidiary particulars is dependent on context. That is, the degree of depth of indwelling differs, depending on the kind of inquiry we are engaged in and its complexity of integration. This degree is proportional to the degree to which the subject matter under investigation lends itself to formalization, exact observation, explanation, and prediction. An exact science like physics requires a low degree of personal participation or indwelling, while indwelling increases in the social sciences, and it becomes more profound and complex in art and religion. Andy F. Sanders aptly employs the thought of Martin Buber to elucidate Polanyi's theory of indwelling: he says that the "more indwelling the understanding of a subject matter requires, the more 'I-Thou' and the less 'I-It' the relation between the knower and what he [*sic*] tries to know [is]."[14]

Polanyi's epistemology forms a seamless garment of knowing. The process of discovery and acquisition of knowledge is the same for all pursuits of knowing. The distinction between science and the humanities is of degree of indwelling.

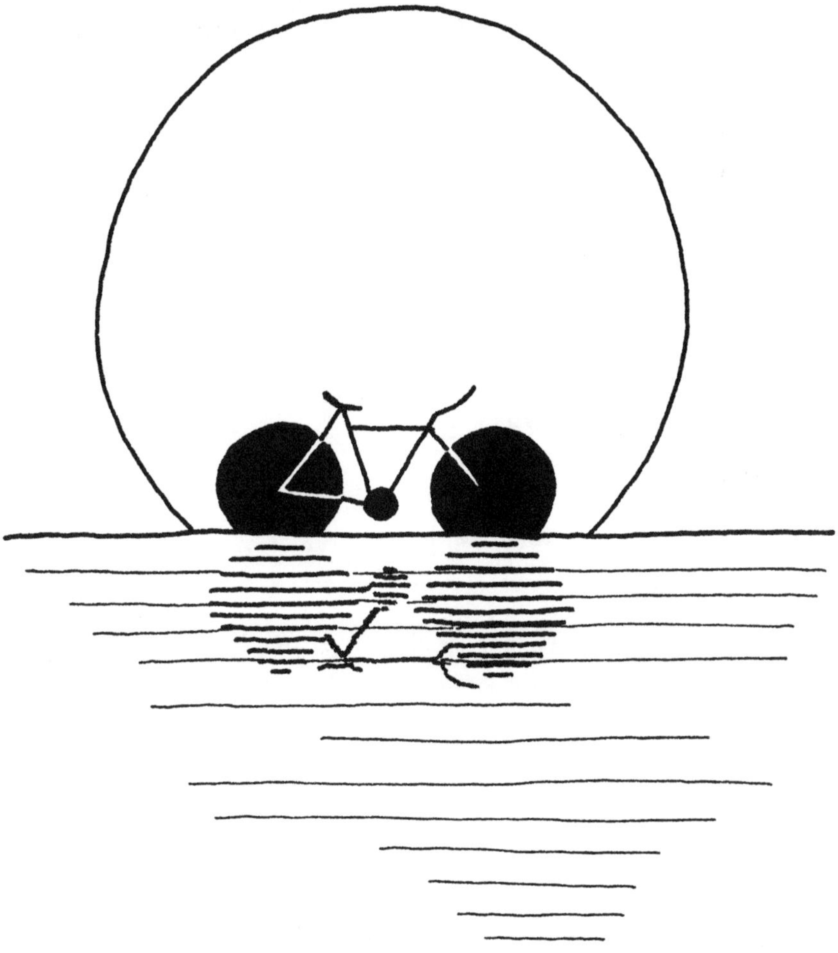

NOTES

1. Michael Polanyi, *PK*, p. 64.

2. Doug Adams and Phil Mullins, "Meaning with the Arts: The Implications of Polanyi's Epistemology for the Arts," *Studia Mystica* 1:2 (Summer 1978): p. 32.

3. Michael Polanyi, "What Is a Painting?" (WP), *The American Scholar* (1970): p. 667.

4. Michael Polanyi and Harry Prosch, *Meaning* (Chicago: University of Chicago Press, 1975), pp. 146-7.

5. Michael Polanyi, *TD*, p. 17.

6. Ibid., p. 18.

7. Lady Drusilla Scott, *Everyman Revived: The Common Sense of Michael Polanyi* (Lewes, Sussex: The Book Guild Ltd., 1985), p. 151.

8. Polanyi, *PK*, p. 197.

9. Ibid.

10. Scott, *Everyman Revived*, p. 187.

11. Polanyi, WP, p. 667.

12. Polanyi and Prosch, *Meaning*, p. 99.

13. Ibid., p. 669.

14. Andy F. Sanders, *Michael Polanyi's Post-Critical Epistemology: A Reconstruction of Some Aspects of "Tacit Knowing"* (Amsterdam: Rodopi, 1988), p. 231.

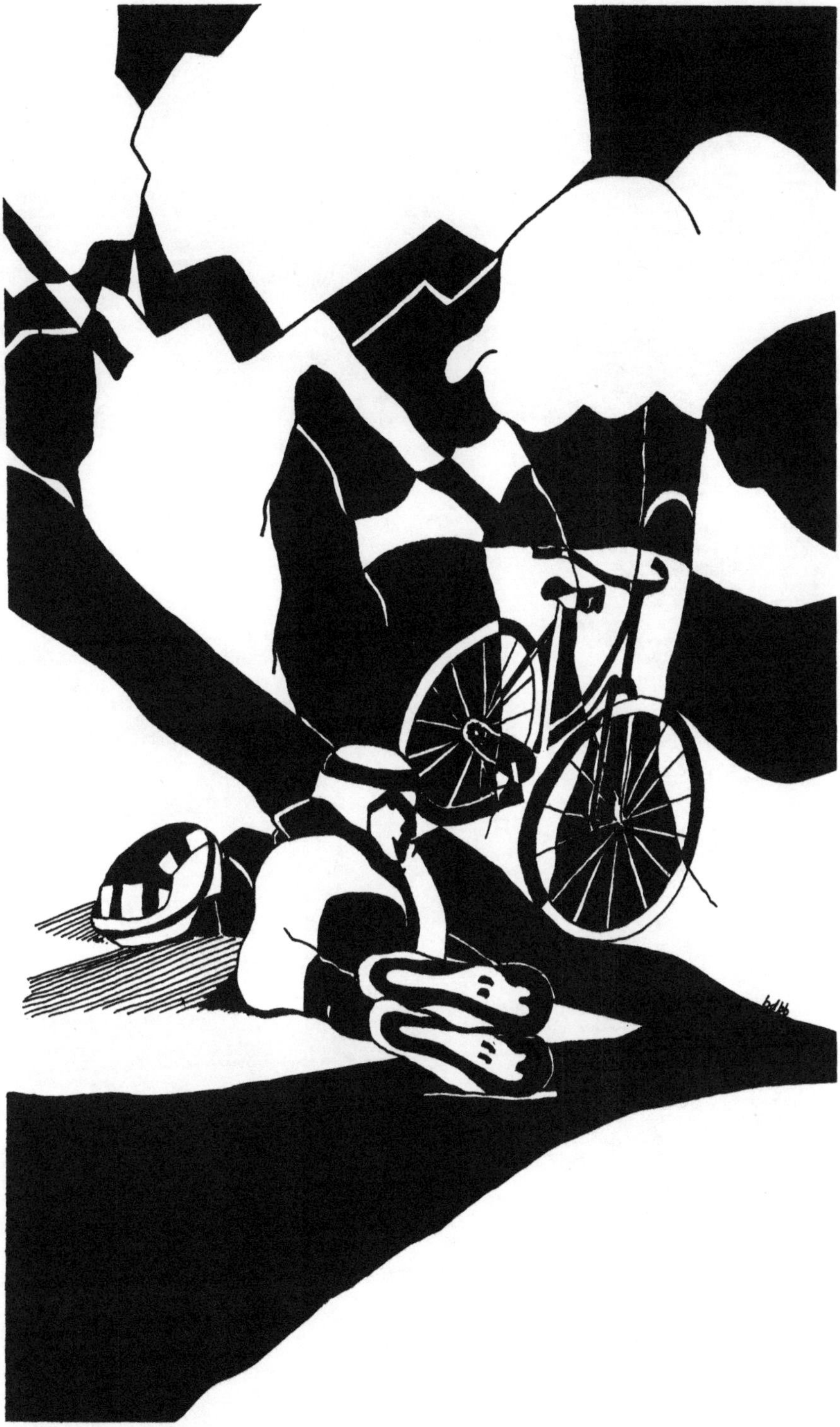

Chapter Three

The Art of Seeing

1. Skills

Skills are, in part, indefinable knowledge that is essential to a task. Polanyi explains that "the aim of a skillful performance is achieved by the observance of a set of rules which are not known as such to the person following them."[1]

While riding a bicycle, a child is unaware of the properties of centrifugal and gravitational forces that s/he actively accounts for by specific turnings of the handlebars and shiftings of body weight. Even if s/he had a scientific knowledge of the rules common to centrifugal force and gravity, the child would still not know how to ride a bicycle. While drawing, an artist's knowledge of exact measurements of objects to be drawn and the distances between them, or the properties of reflective light upon the given objects, does not produce artistic drawing. The rules, in these cases, may be helpful guides toward a deeper knowledge of these skills, but they do not determine their practice.

The rules of such performances, bicycling and drawing, are valuable only when integrated by an individual who already possesses a practical knowledge of the art. A rule, when known, is something that we focus on when we are learning it, but in order for it to be practiced or helpful, we must integrate the rule into our tacitly held knowledge. Skillful performance requires that we focus on a task, not its maxims. A bicycler may gain greater insight and guidance in the art of bicycling through a knowledge of its maxims, but the maxims cannot replace nor establish the riding skill. The maxims of drawing may offer valuable guidance to the artist, but they cannot fully account for the artist's skill.

Therefore, since an art cannot be specified in detail, it cannot be taught through its maxims. It can only be taught through personal guidance. For example, a parent does not give the child a science lesson in preparation for the first bicycle ride. S/he usually demonstrates riding the bicycle, then lifts the child on to the bicycle, and gently helps to physically guide the child's ride until s/he feels the child's balance. At that time, the parent lets go and hopes for the best.

Learning to ride a bicycle is a time of physical trial and error where few words are shared. Indeed, words can get in the way of the learning process. If the parent says, "Okay, I'm going to let go now, keep your balance," before letting go of the bicycle, the child will be unsure of what "balance" means, lose confidence, and most likely fall. By bringing the idea of balance into the discursive mode, the child seeks for a discursive rule or maxim for balance. Since none is available to the child, the child suddenly feels that s/he does not know what balance means, hence loses a grasp of the physical unspecifiable reality of balance, and therefore falls.

If the parent, upon feeling the child's balance, simply lets go without a word, the child usually rides a few moments successfully. Often, the child will then state, "okay, let go," and the parent joyfully responds, "I already did!" The child was able to ride the bicycle without expressly knowing it. An indefinable passing of knowledge occurred between parent and child.

Nor does the art teacher offer to the student a set prescription on how to draw. "How to draw" books that attempt a detailed prescription for drawing may succeed in teaching the student a mechanical knowledge of drawing specific objects in ways that result in impersonal, illustrative drawings. Such drawings are not considered art.

I recently ran an experiment in a weekly children's art class. On a normal class day, I insist the children draw from life. I set up a still life, bring in a pet, take them outside, or arrange other life presentations. The children immediately become engaged in their work and create wonderful, fresh, and lively drawings. In the experiment, however, I told the children we were going to draw birds and offered them books to follow on "how to draw birds."

A few children were surprised by my offering, others were quite happy to have something to copy. During the work, the children became restless. Their drawings were stilted, and I found myself floundering with disappointment in response to them. The drawings were mechanically successful but lacked the depth and life of art. An expression of my dissatisfaction would not make sense to the children since they followed directions.

Most children were comfortable doing something that did not demand creativity. They were in control of the product rather than being challenged yet free to risk a personal form on that white paper. They allowed the rules on drawing birds to determine their art. The result was not art but copies of the rules. The "how to draw" rules could not replace their working knowledge of drawing.

The experience was not, however, a complete loss. Those children of exceptional creativity found expression through playing with the medium used to copy the bird forms. And, in our continued study of birds, I did not allow the "how to" books, but the children remembered those books and what they had learned. They integrated the "how to" knowledge with their practical knowledge of the art of drawing. The rules became guides that no longer impeded creativity but deepened it.

The art teacher, rather than offering rules on how to draw, often finds herself/himself at a loss for words and teaches drawing through example and critique of the student's work. I often find myself drawing on the student's drawing, not to improve the drawing, but because it is the *only* way I can possibly relate to the child a skill that needs to be learned. Words are inadequate, action is everything. Yet the task of the teacher is not to make a drawing for the child to copy; the results of this method may seem on paper to instruct drawing, but in actuality the child is not challenged to "see," to integrate the specifiable maxims of drawing with the practical unspecifiable skill of drawing.

The teacher's task is to excite the child with a problem, and then gently guide the child to ferret out for himself/herself the way to draw the subject at hand. I find words, then, to be most meaningful at the end of the day when we hang up the drawings for critique. In the critique the children are asked to notice how rich a certain black is, or how sensitive a line is drawn, or how the energy flows. The words of critique again are not specifying rules, but bringing to consciousness what works and does not work within a drawing.

The articulate aspects of drawing, such as illustration, principles of light and dark, perspective, may be learned through rules taught in the classroom or found in books, but the unspecifiable skill needed for the creation of an artistic drawing is bound to the teacher-student relationship.

The drawing teacher sets up a still life, not for the student to look at in itself, but as a clue to the unknown drawing. The still life functions as a pointer to, and as a part of, the unknown drawing. The problem of creating art through the data presented by the still life cannot be solved by following a specific procedure (such as found in "how to draw" books), nor by following definite rules (common to volume, perspective, or design theory). The teacher is called to present a problem that puzzles or worries the student; the student is called, through a skill that is largely unspecifiable, to discover a solution that relieves the burden of the problem.

In presenting a problem that anticipates the unknown drawing to the student, the teacher must be aware of the tension between the student's ability and the difficulty of the task. If the task is too difficult for the student, the problem will be insoluble and the effort wasteful and frustrating; if too easy, the results will be mediocre. It is a skilled teacher who can successfully judge the suitability of a problem for a given student.

Creating art, like any discovery, is not a logical exercise, but the overcoming of an obstacle that Polanyi calls a "logical gap." He says that, "to speak of the width of the logical gap [is to] measure the ingenuity required for solving the problem."[2] Through ingenuity, the discoverer plunges across the logical gap to gain a foothold at another shore of reality.

So, the art student is encouraged to integrate articulate maxims of art with unspecified skill in drawing to present a distinctive drawing of still life that deviates in a surprising way from the actual data of the objects assembled. When the maxims are incorporated they are interpreted by the student to fit and guide her/his style. The student solves the problem of an artistic drawing of the still life through an heuristic process which leaps across a logical gap to create something new.

Learning to draw, like all skills, relies on the practice of drawing under a teacher. Drawing cannot be learned without working out concrete problems of drawing. But since the ability to recognize and initiate a problem, an unknown which lies across the logical gap, is a skill in itself, the teacher is essential to the learning process. An accomplished artist ready to leave her/his teacher's side has above all the ability and desire to discern and solve an artistic problem.

The necessity of a teacher is profound and is made evident in that an art no longer practiced for a generation is lost altogether. Polanyi cites our inability to reproduce the Stradivarius violin which was produced more than 200 years ago. The art of making this violin has been lost despite what we know through microscopic, chemical, mathematical, and electronic analysis of this instrument.[3]

Lost to the American culture is the quilting bee, the gathering and making of community around a quilt in progress. The resurgence of quilting occurred just as the last generation of quilters was dying out, and our contact with these quilters has saved quilting from hopeless loss as well. We who have stitched the art of quilting back to life explain our interest and knowledge, "my grandmother quilted." Contemporary quilters witnessed women quilting in their last days. These women were often too old to get out and quilted at home or in an institution as a way to pass the days just before death. The generation of the quilting bee had passed.

At the quilting bee, the piecing and quilting of patches was a stitching together of stories into patterns that represented concerns and life experiences. The patches represented people, moments, and occasions; when sewn together, the individual stories were integrated into a larger story. A communal story of life: of its joys and sorrows; of its everyday routine and special activities; of its passages and heritage. Quilters stitched together a narrative that interpreted and formed their history which, in turn, gave them a sense of self and community. As the quilt took shape, so did lives.

The development of fabric shapes and patterns offered new communal understandings for the pioneers and their families through the generations that participated in the bee. Innumerable books have been published about the bee and the lore behind the quilts, but all of our historical knowledge cannot revive the life of the bee's existence. The American quilting bee is of ages past, lost to our contemporary culture.

Knowledge is passed from one generation to the next via tradition. The unspecifiable elements of knowledge possessed by one generation are received by its successors through mindbodily imitation and attention. We recognize authority in order to learn a traditional art, but "traditionalism itself is merely a rule of art which, like all such rules, becomes absurd when taken out of context and accepted as a precise prescription."[4] Traditionalism, or the valuing of tradition as always correct and an end in itself, does not foster knowledge and its skillful application.

Practical knowledge is more truly embodied in the action of the master than it could ever be expressed in maxims. In learning a skill the student places trust in the teacher and acknowledges his/her authority. In doing so the student assimilates consciously and unconsciously the rules of the art, rules that are explicitly known and rules that are not known to either the teacher or the student. Whereas artistic technique can be systematically specified and learned impersonally, artistic content cannot.

After having built a foundation of knowledge, through instruction, modeling, and practice, the student exercises personal judgment and creative power to understand, apply, and break out from the maxims, including tradition. Otherwise, the maxims fall into absurdity, tradition becomes authoritarian, and the art is empty. It becomes form without content, not art. When tradition and maxims are integrated into the artist's skillful performance of creation, the artist becomes a living part of the artistic tradition.

A skill, however, is a motoric practical form of largely unspecifiable knowledge. To understand practical human knowledge that cannot be formalized but instead is unspecifiable, is to understand skills. Skills are motoric in that an intense personal participation is always embodied in a skillful performance, be it the large motoric performance needed to ride a bicycle or the more subtle motoric performance demanded in the perception and eye-hand co-ordination of the visual artist. One practical form of knowledge is of a means-end type. Animals can acquire practical knowledge, and we can formalize the operative principles by which they apply it. When an animal learns to press a lever at certain intervals in order to release a pellet of food, the animal is said to be practicing the means-end type of knowledge.

According to Polanyi, skills may be understood in the wider sense to "include all motoric performances of an unspecifiable kind and [we] may note that some skillful motoricity enters into every kind of intelligent performance."[5] And in the narrower sense, a skill may be defined as "a pattern of motoric behavior carried on by conscious effort and, in its most characteristic instances, guided by strained attention towards an aim well known to the performer, though he may not be able to designate it clearly in words."[6]

The bicycler aims at keeping the bicycle in balance while moving forward, but the aim of a visual artist is to discern and solve an artistic problem which is not well known, but which anticipates the unknown. Therefore, the task of the artist is more a judgment as exercised by the connoisseur, which appreciates a complex relationship between unspecifiable elements including skills.

2. Connoisseurship

All human knowledge is on a connoisseurship continuum. Connoisseurship is a theoretical type of knowledge which includes the sign-event and alternative part-whole types of knowledge. When an animal learns that the ringing of a bell signals the approach of food, then the animal has learned sign-event knowledge. Or when the animal learns a maze via the shortest route, then upon discovering the shortest route blocked, it learns an alternative shortest route, the animal has achieved alternative part-whole knowledge. Both of these cases can be formalized as a formula (when p, then q), in the maze experiment, or through the drawing of a map which shows the alternative routes. But when types of theoretical knowledge cannot be specified, as in human knowing, it is connoisseurship.

Connoisseurship can be said to be more physically detached than the practical motoric skills. Yet we cannot refer to skills without practicing connoisseurship. Since skills arrange unspecifiable elements for the achievement of an aim, and we assimilate those elements because we appreciate them, the accomplishment of a skill is also a feat of connoisseurship. As we exercise skills, so must we also practice connoisseurship.[7]

While skills and connoisseurship may be partly described, separately, as they are based on different types of knowing, they are inextricably bound together. We cannot practice a skillful action without an appreciation of its aim, nor can we practice judgment which appreciates complex relationships, without exercising the appropriate skills to the given task. The relationship of skillful action to skillful knowing is of degree. The bicycler, who appreciates the art of bicycling, is more intensely involved in the motoric demand essential to the skillful performance of riding a bicycle. The visual artist is more concentrated on the judgment of the complex relationships that point to an unknown than on the skillful performance required in the achievement of the artwork. Since connoisseurship is as much the art of doing, as it is knowing, let us consider connoisseurship more thoroughly.

All that has been stated thus far regarding skills applies equally to connoisseurship. Connoisseurship, like skills, is largely unspecifiable and can only be learned by example not by precept. A wine's taste cannot be scientifically measured. A master's guidance while actually tasting wine is how wine connoisseurship is achieved. On many occasions in medical science diagnosis is dependent on an unspecifiable skill that can be learned only through a long course of experience under the guidance of a master.

Twenty years ago, babies rarely survived a particular respiratory condition. With modern technology and recent medical discoveries, a surgery has been developed that allows for the survival of these babies. My husband and I adopted such a child. I will illustrate the process of learning connoisseurship through my personal experience with this son.

The first physician who attended to our new baby, then seven months old, upon seeing him responded, "I've had one of these children in my practice before. They are sick all the time. I finally have the other boy, now twelve years, to where he only gets sick once or twice a year." Our baby, whether well or sick, had a quite noisy and abnormal breath. My question to the physician was to inquire how I would know when the baby was sick since he sounded ill all the time. "Oh, you'll know. You will be able to tell the difference between a well breath and a sick one." I was looking for maxims to guide my way in caring for this child. Instead, I was offered the opportunity to train my ear to the child's breath, to become a connoisseur of respiratory ailments. I did not know the difference between the well and sick breathing patterns at first. But through daily experience over time, under the guidance of the physician, I learned to recognize when my baby was sick. The physician's skilled senses and use of sensitive medical equipment taught me how to detect illness in my son.

When our son was two years old, we moved to another part of the state. The baby's new physician had never experienced a child with his condition. Fortunately, the physician was committed to the child's care and learned all he could to help the child.

Over time, I trained the physician. I had developed an ear for the child's health and found that I needed to share that knowledge with the physician. Maxims, such as the presence of fever, cough, sluggishness, or sputum, were not useful in our son's case.

At one point, our son was seen by the chief pediatric pulmonologist at a prominent medical school. This physician could find no explicit evidence for our son's condition. He emphatically stated to me, "There is nothing wrong with this child. Look at how stocky he is. Children with trachea malacia (one aspect of his condition) are thin and pale. Your child is far from thin and pale. I guarantee you that this child does not have trachea malacia." I knew that the physician was wrong, but I had no explicit evidence to disprove his opinion.

Our son's primary physician trusted my judgment and sent us across the country to see another prominent pediatric pulmonologist who used hypnosis for diagnosis. He has found that hypnosis does not impair the normal workings of the body, and therefore, a more accurate diagnosis can be made on difficult cases such as our son's.

Upon initial examination, again, there was no evidence of "sickness." Yet when our son was hypnotized, and the physician went inside the body to examine the respiratory system, my suspicions were confirmed: the child had a severe trachea malacia and raging infection.

Connoisseurship, the ability to judge a pattern of unspecifiable elements, is the only explanation of how I knew the child was not well. A video tape was made of our son's condition which we then showed to the other pulmonologist, not to spite him, but in hope of helping him realize that diagnosis cannot rely on maxims alone. In our son's case, measurable grading could not replace the unspecifiable skills and connoisseurship that can only be passed from master to apprentice.

Polanyi points out that a case like our son's is not the exception but the norm. The art of knowing and diagnosis has remained unspecifiable and can only be communicated by example. This actuality can be demonstrated at the very heart of science.[8]

The same can be said for the creation of art. Incarnating an as yet-unknown artwork on a piece of paper or canvas, for example, relies on a learned capability to judge and appreciate specifiable and unspecifiable elements that point beyond themselves toward a new comprehensive whole. While skillfully making art, the artist is able to perceive through an unspecifiable sensitivity, the correctness of color, form, line, design, etc., essential to discovering and completing the aimed for but as yet-unknown artwork.

Instrumental elements which are largely unspecifiable in this skillful exercise function as the elements of an observed comprehensive whole in the practice of connoisseurship. Both the art of doing (skills) and the art of knowing (connoisseurship) are aspects of the act of extending our person into elements that are essential to the achievement of an aim and the appreciation of a comprehensive whole. It is the skilled artist as connoisseur that is able to create works of art.

Skills and connoisseurship are cumulative achievements that depend on innate faculties and personal traditions. They are learned through dwelling in and trusting a community and its tradition. The skills acquired from others allow us to break out toward new more complex achievements. The framework of learned skills and judgments which others had before us serves as a foundation upon which we build our own mastery. We integrate new and acquired skills into a tacitly held network of intricate and novel achievements. This network is an active component of our encounter in our world.

A skill that is tacitly rooted in the act of art appreciation is aesthetic seeing. Just as an intentional and existential commitment enters into the process of making art, so the viewer submits to an artwork committed mindbodily, open to discovering its meaning. The viewer participates in an artistic event distinct from ordinary experience. S/he does not mistake a painting of a landscape for a real landscape nor a murder in a play for a real murder. Polanyi characterizes this as the process of symbolization in which the beholder is carried into an imaginative integration so that s/he may better perceive ordinary experience. Because art is not imitative of natural reality but creates something new which exceeds the vividness found in the everyday, art requires imaginative seeing.

Aesthetic seeing and art appreciation are being lost to our contemporary American culture. Technology, mass media, and a reliance upon discursive expression have reduced our ability to see aesthetically. Contemporary viewers of art have become aesthetically calloused. The underlying and contemporary problem of seeing must be addressed if a relationship between art and religion is to be discovered, and if visual art is to be understood as theology. Before the theological dimension of art can be discovered, and the theological sensibilities of the artist, the religious community too, must develop art responsiveness. They need to develop the skill of seeing imaginatively.

3. Perception and Imagination

Visual perception consists in the integration of clues evoked in our mindbody by the impact of light to produce the sight of their joint meaning. Clues include the retinal image, the feeling of eye muscles, the tension of bodily muscles supporting the position of the head, the sensations from the inner ear, the hidden memories which shape interpretation of objects, as well as the neural traces evoked by the light reaching the eyes. Perception is grounded in this integration of clues evoked in our mindbodiliness which makes sense of the objects presented to our vision. The clues of visual perception are relied upon subsidiarily in order to integrate them to their joint meaning, the sight of the object.

Perception is not instantaneous, nor is it effortless. The simplest scanning of a scene requires time and a remarkable feat of integration. Our eyes see the objects before them in a series of snapshots taken from consecutive positions at the rate of about three to four shots per second. In a period of ten seconds spent looking at an object, thirty to forty different snapshots are collected. This collection is then integrated into a single view of the object looked upon. Perception enables a single sight to merge from the whole collection of consecutive pictures.

The collection of pictures is not random. We select the details that are of the most interest. Perceptual integration quickly establishes what is being seen, and the eyes focus on what are deemed to be the most important aspects of the scene before us.[9]

The effort of the eye to present an image of clarity is the result of self-set perceptual standards. The muscles of the eye adjust the thickness of its lens, and the collection of images are selected and integrated in accordance with a learned interpretive framework. When a baby moves a rattle toward its eyes and away again, the baby's perception must choose between seeing the rattle swell up and shrink alternately, or see it change its distance while retaining its size. The baby chooses the second alternative. This choice guides our perception into a universal interpretive framework which sees objects as retaining their size and shape when seen at different distances and angles, and their color and brightness when seen under different types of light.[10]

Even when retinal evidence is to the contrary, perceptual adjustments are made to maintain the standards of correct seeing. The subsidiary clues of visual perception work toward a coherence of the world seen. The process of integrating visual clues into their joint meaning is a lifelong course in application of and reliance on established sensory clues.

Making and appreciating art require the artist and beholder to acquire a new way of seeing things that re-integrates their established perceptual interpretive framework. Learning "to see" imaginatively is a breaking out from the perceptual standards in which we dwell. The skill of seeing, learned by repeated seeing and in no other way, stretches, forms, and filters our horizons into a discriminating perception that is more than itself.[11] The artist and viewer, rather than relying on what they have learned to see in the past, learn to see in a new way.

For example, a student needs to learn, when setting out to draw a person lying horizontally with feet close to the student and the head at a distance, to see the body differently than it is ordinarily seen. The student learns to see the feet swelled and the head shrunk rather than seeing the feet closer and the head at a distance. The opposite choice of the baby rattle experience is assumed: variances in seeing are realized rather than consistency. The feet are drawn proportionately larger than the whole of the rest of the body; the body is drawn progressively smaller as it moves toward the head. In art this is called foreshortening. What is closer is drawn larger; what is at a distance is drawn smaller. Objects, in the artist's eye, do not retain their size, but change when seen at different angles and distances.

The same is true for color. The artist learns to see variances in color rather than consistency due to changes of illumination. This breaking out from existing perceptual frameworks demonstrates our ability to intentionally prompt perception to explore and assess perceptual clues in new ways in the quest of discovery.

Most perceptual operations are achieved tacitly; with no explicit awareness of and no direct control over them. This is akin to the skill of riding a bicycle. The precise motoric operations are largely unspecifiable. Even if they are known, they are not controlled directly. Motoric feats are accomplished by general intention or purpose not direct control of elements. Our intention to move our legs in a circular action sets in motion a complex bodily mechanism in its service. So too, our intention to see an object sets in motion the visual mechanism of our mindbody in the service of its purpose.

The motoric action needed to ride a bicycle requires more effort than the simple scanning of a scene. But when the light is dim, perception is more strenuous and we send out greater thrusts of the imagination. So too, in the effort to break out from existing frameworks, the perceptual feat is more demanding. Yet in all cases, perception works within our imagination. Polanyi calls "all thoughts of things that are not present, or not yet present—or perhaps never to be present—acts of the imagination."[12]

The lifting of our arm, riding a bicycle, perceiving are all intentional acts of the imagination. The more challenging the feat of seeing, the more strongly the imagination is invoked. A gap between our intention to sight an object and the achievement of that sighting arises when our aim meets with difficulty. Greater thrusts of the imagination are required to close the gap between a complex intention and its performance.[13] The imagination is able to bring together subsidiaries that we have been unable to integrate in a difficult feat.

Visual perception is a purposive action and requires some thrust of our imagination. Perception, like all action in progress, aims at the future, and we can only attend to the future in our imagination. The same operation is at work in motoric action and visual perception. The imagination activates mindbodily mechanisms over which we have no control and thereby implements our intention.

Once sight has been achieved, argues William H. Poteat, anticipation ceases. "Even in a novel sight, according to our picture, there is nothing *before the eyes* to anticipate. It's all there at once in a slice of dead (visual) space."[14] The particulars which are integrated appear simultaneously co-present with one another; that is they seem to be before our eyes in the same instant.

Visual perception is immediate when compared to other modes of perception in which the temporal is clearly present. In speech, for example, the tacit components are not simultaneously co-present; they are separated by more noticeable increments of time. The words of speech occur temporally one after another. Perception, in Poteat's logic, is static, finite, determinate, self-contained, and complete in itself.[15] Given the collection of three to four snapshots per second which are integrated into a single view of an object, vision does not appear to require temporal thickness in its achievement. It is an almost instantaneous act characterized as a dead slice of visual space.

We cannot in actuality remove our mindbodies out of the temporal thickness of experience, and thus we do not fully experience feats of seeing as static. However, the co-presence of particulars before the eye suggests that visual acts of perception are atemporal, or finite and static. Speech, on the other hand, is described from our experience of its temporal thickness or sequential quality and thus is characteristically dynamic. Vision, seemingly separable from the temporal dimension of our being in the world, appears to endure through time unchanging; it is "indistinguishable from eternity."

A painting, for Poteat, is an object of sight that seizes, arrests, and holds the eye. It embodies a rich slice of dead visual space. The artwork presents before the eye a spatio-temporal configuration which integrates a far more complex object of vision than ordinarily perceived. The visual feats required to produce the joint sight of the artwork's particulars can only be achieved temporally.[16] Art is experienced temporally through our mindbodily being in the world. Works of art present to our vision a "moment of form" which is not easily perceived. We send out thrusts of our imagination to explore art through noticeable periods of time.

Acts of ordinary visual perception integrate particulars which are temporally close. Because such acts of integration are habituated, we have a sense in seeing that the particulars are simultaneously co-present before the eye in a slice of dead visual space. A scene, once sighted, does not seem to require anticipation to be held before the eyes compared to acts of listening which include anticipation throughout the hearing of the speech. Although ordinary feats of seeing are not temporally thick and therefore require little imagination, acts of artistic visual perception are temporally thick and so compel great imagination.

The elements of visual art are a complex configuration. While they appear as present before our eyes in an instant, we cannot attend to them in that instant. We must intend to sight the motifs presented continuously as we stand before the artifact. The creative imagination is invoked to expand our anticipation of meaning to its discovery. The sighting of an artwork is never complete. We persist in creative anticipation of the artifact's content in the duration of the time it is before our eyes.

Perception of visual art continuously requires thrusts of our anticipatory and creative imagination. The artwork embodies a "moment of form" which is co-present before the eyes. Its material form endures unchangingly over time. Our apprehension of the artwork occurs temporally *ad infinitum*. This is true even for the artist. For example, in 1976, I created a drypoint entitled *Dance of the Twigs* (Plate 2) which expressed a celebration of life. Over time, however, I came to discover that the images were not joyful but images of pain, anger, and domination. I became aware that a perceptual tension is present in the artwork between its material finitude and its inexhaustible potential to progressively reveal meaning. The artifact may appear to present itself to the eye simultaneously, but its meaning cannot be grasped in that seeming instant.

Visual art's meaning is temporally thick when compared to ordinary perceptual meaning. Art is not static, finite, nor determinate; it is dynamically finite and indeterminate. Art incarnates a *live* slice of visual space. Its appreciation requires great thrusts of the imagination. Polanyi's description of what the imagination is helps us to understand the imagination's role in beholding art as well as what art presents.

Imagination "is the nucleus of all discovery, invention and artistic creativity."[17] The integrative powers are largely spontaneous; thus, Polanyi names them "intuition." The thrusts of the imagination propel and supply intuition with suitable material that leads toward discovery. Imaginative effort and intuitive observation are merged to account for all apprehensions of meaning.

Intuition sparks the imagination which implements our intentions; imagination imbued with intuition informs and enriches our tacit mindbodily being. Discovery goes like this: first intuition offers an idea to the imagination to be considered; then, intuition guides the imagination's ferreting out of possible clues; and finally, the imagination points to a solution that arises intuitively.[18]

The work of intuition has three aspects. (1) Dynamic intuition senses clues which lead to a hunch that suggests a deeper but as yet-unknown coherence of reality. (2) Strategic intuition then serves as a skillful guide that recognizes the value of a problem and senses clues to its solution. (3) Creative intuition instinctively realizes the concluding idea. It is the spontaneous "Aha!" of discovery. The creative intuition responds to the imagination as do the reflexive movements of our muscles when we intend to move our bodies. Intuition searches for a good problem, estimates its significance, and then presents its solution.[19]

Discovery is not achieved solely through spontaneous sensings of our intuition. Our intuitive faculties are empowered by the imagination. The imagination thrusts forward with a deliberate intention, and intuition integrates the clues that imagination lights upon. Intuition works on a subsidiary level, sensing and integrating clues that are largely unspecifiable, while the deliberate focal acts of the imagination seek reality and its meaning. "Intuition informs the imagination which in turn releases the powers of intuition....[When discovery is achieved] intuition recognizes our final result to be valid, and our imagination points to the inexhaustible future manifestations of it."[20]

Our use of language can serve as an example of how we produce and discover meaning through intuition's guidance of the imagination. When we name something, a relationship is established between the word used and the thing. The meaning of the word is enlarged to include this new relation as well as a whole range of instances in which the word was used before. Words are steadily modified as they are used over time; thus, a word used by one generation may have a changed meaning when used by the next generation. We are able to express what we mean by using that word, or to understand someone else's use of the word, by an effort of our searching imagination and intuition's spontaneous recognition of its appropriateness.

Our intention to express meaning or understand what someone means sets in motion our imagination, which aims at successfully producing or comprehending meaning. The intuition senses the presence of meaning from the word uttered; then it guides the imagination to its set of clues and subsequent discovery. This process, which includes an alternation between an analysis of its parts and an integration of the parts to the focal whole, is fundamental to all mindbodily knowing, be it theoretical or practical, tacit or articulate.

To summarize, all meaning is achieved through the same implementation of the imagination, but the thrusts of the imagination required to perceive a simple scene, name an object, or make a painting are not the same. The imaginative performance required to create a painting is much greater than is required to scan a scene. The complexity of intuitive integration bears directly on the measure of imaginative effort. A highly imaginative person is able to perform complex integrations that a less imaginative person may be unable to achieve. The integrations required to ride a bicycle are much less complex than are those needed to make a scientific discovery or a painting.

An easily apprehended coherence requires less indwelling than a complex one. Hence, the greater the complexity, and the greater the indwelling, the greater the imaginative performance. A scientist seeking a discovery carries out a highly imaginative integration while dwelling deeply in the clues that lead to that discovery. Artists while painting, too, have vast demands put upon their imaginative powers as they dwell in the depths of their world in order to break out from it. The scientific discovery and the painting are both achieved by great thrusts of the imagination. But once the scientific discovery has been made, little imagination, little indwelling is required for others to possess the discovery. Therefore, an exact science like physics requires less personal participation than a less formalized field such as art.

The case is different in the arts. An artwork to be apprehended requires large imaginative effort on the part of the viewer. The beholder dwells passionately in the artwork as a seeker employing exceptional thrusts of the imagination in order to perceive its meaning. The intuition functions to incite the imagination's quest for meaning; art functions to incite the imagination of its public and its quest for communal meaning. While our possessing of scientific knowledge does not demand the same integrative powers as does scientific discovery, our understanding of art does require a much richer measure of imaginative integration akin to that of artistic (and scientific) discovery. The beholding of art is a type of discovery.

Polanyi has stated "that a work of art represents *facts of the imagination*, it does not affirm any *facts of experience*....To appreciate a work of art requires that we believe in that which it means, not a disbelief in something that it does not and must not assert."[21]

A fact is an observed occasion that bears on reality and its meaning. A fact that does not bear on reality is meaningless; meaning that is not supported by fact is powerless. Facts are instances of natural reality that make sense of the clues offered by experience to our mindbodies. Reality is what manifests itself indeterminately in the future. Since a fact is a known aspect of natural reality it too may be indeterminate in the future. Furthermore, a fact can only be known personally, through our mindbodily being in the world. The evidence that supports a fact is personal evidence submitted with universal intent but also fraught with the possibility of error.

The establishment of a fact can be most difficult. Reflection is necessary on the underlying assumptions that lead to its assertion. This is a challenge because some of the underlying assumptions are unspecifiable and cannot be sufficiently analyzed. Even a scientific allegation of fact relies on the unformalized tacit dimension that bears upon all our experience.

Ultimately, an organized consensus that we, the community-at-large, recognize as an authority validates the fact by stating that we believe a fact to be true given the evidence. At the same time, the accredited community recognizes that new evidence could reveal or even has revealed the error of the alleged fact and a new fact could or should be submitted in its place.

Facts of experience and facts of the imagination are made up of varying degrees of specifiable and unspecifiable components. Facts of the imagination tend to be less specifiable and thus are more difficult to verify. Indeed, Polanyi asserts that art cannot be verified, only validated.

Verification is an empirical endeavor of observation and repeatable experimentation that applies primarily to science and not other systems, such as mathematics, religion, and the arts. Like verification, validation relies on commitment but it requires a greater degree of personal participation. Both verification and validation are formally indeterminate procedures that require personal participation and judgment and are subject to authoritative consensus and future discoveries; they claim knowledge of something real and outside of the knower.[22]

Validation relies more on internal rather than external evidence; it requires a high degree of imaginative effort. But like verification it affirms assertions as meaningful interpretations of our experience of reality. These personal affirmations necessitate that we proceed with universal intent lest validation becomes a subjective endeavor. Validation is subject to the claims of experienced reality. It is not arbitrary or self-willed. The recognition of truth, the assertions of experiential and imaginative facts, hinge on the rightness of our mental acceptance. Our tacit knowledge and our experience of reality jointly guide all of our intellectual achievements.[23]

Art is an "event" because art is significant, it is a novel occurrence, it requires a high degree of personal participation to be grasped, its meaning is grasped progressively as we attend to it mindbodily over time, and it is public. The factuality of art is its presentation of an authentic event of supra-natural reality that makes sense of a complex set of clues grasped in an extraordinary way. Aesthetic form must be true in order to be meaningful, and when validated it can be said to be a "fact" because it is an observable event that bears on reality and its deeper meaning. When it is validated in the public domain it guides our thoughts throughout our culture.[24]

Being a fact, art is indeterminate in the future. As a fact of the imagination, it concerns itself with future manifestations of reality that are inarticulate. Thus, we cannot finally and conclusively state its meaning. New evidence, subject to our experience in the world and with the artifact, will affect our understanding of art's meaning.

Our experience of the artwork is steadily modified in several ways. First, the actual comprehensive artifact changes us and our successive understanding of it. Second, the clarifying analysis of the artwork enlarges our apprehension of its meaning. Third, who we are as a person progressively changes and thus so do the frameworks that we bring to bear on the artwork. Fourth, each generation will bring its own socio-historical context to bear upon it. However, art appreciation is not in service to the subjective whim of the beholder. Art is more than an embodiment of an authentic subjective experience. Art represents what quickens our intentions as we make contact with reality.

A bicycler imaged in art serves as a motif through which the artist expresses an experience of supra-natural meaning. In so doing the artist embodies a slice of supra-natural reality and presents an inarticulate aspect of reality. S/he incarnates beauty through which the viewer is invited to participate in the artifact and grasp its meaning.

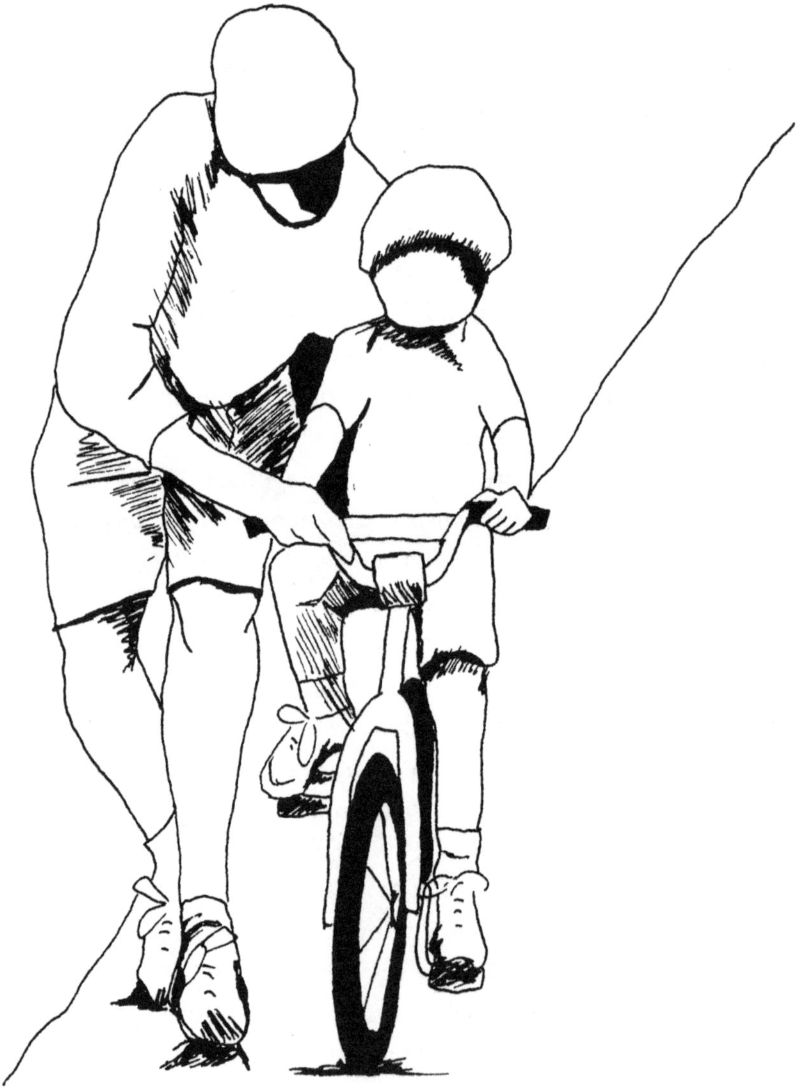

NOTES

1. Michael Polanyi, *PK*, p. 49.
2. Michael Polanyi, "Problem Solving," *The British Journal for the Philosophy of Science* 8 (August 1957): p. 93.
4. Polanyi, *PK*, p. 53.
5. Michael Polanyi, "Skills and Connoisseurship," *Atti Del Congresse di Metodologia* (Torino, 17-20 December 1952): p. 389.
6. Ibid., p. 382.
7. Ibid.
8. Ibid.
9. Polanyi, *PK*, pp. 54-5.
10. Michael Polanyi, "From Perception to Metaphor," in M:P, p. 6.
11. Polanyi, *PK*, pp. 96-97.
12. John Dillenberger, *A Theology of Artistic Sensibilities: The Visual Arts and the Church* (New York: Crossroad, 1986), p. 244.
13. Michael Polanyi, CI, p. 100.
14. Ibid., pp. 100-101.
15. William H. Poteat, *Polanyian Meditations: In Search of a Post-Critical Logic* (Durham, N.C.: Duke University Press, 1985), p. 84.
16. Ibid., p. 74.
17. Ibid., pp. 60-1.
18. Polanyi, "From Metaphor to Perception," in M:P, p. 5.
19. Polanyi, "Works of Art," in M:P, p. 18.
20. Polanyi, CI, pp. 98-104.
21. Ibid., p. 104.
22. Polanyi, "Works of Art," in M:P, p. 14.

23. Polanyi, *PK*, p. 202.

24. Ibid., *PK*, p. 321.

25. Polanyi, "Visionary Art," in M:P, p. 10.

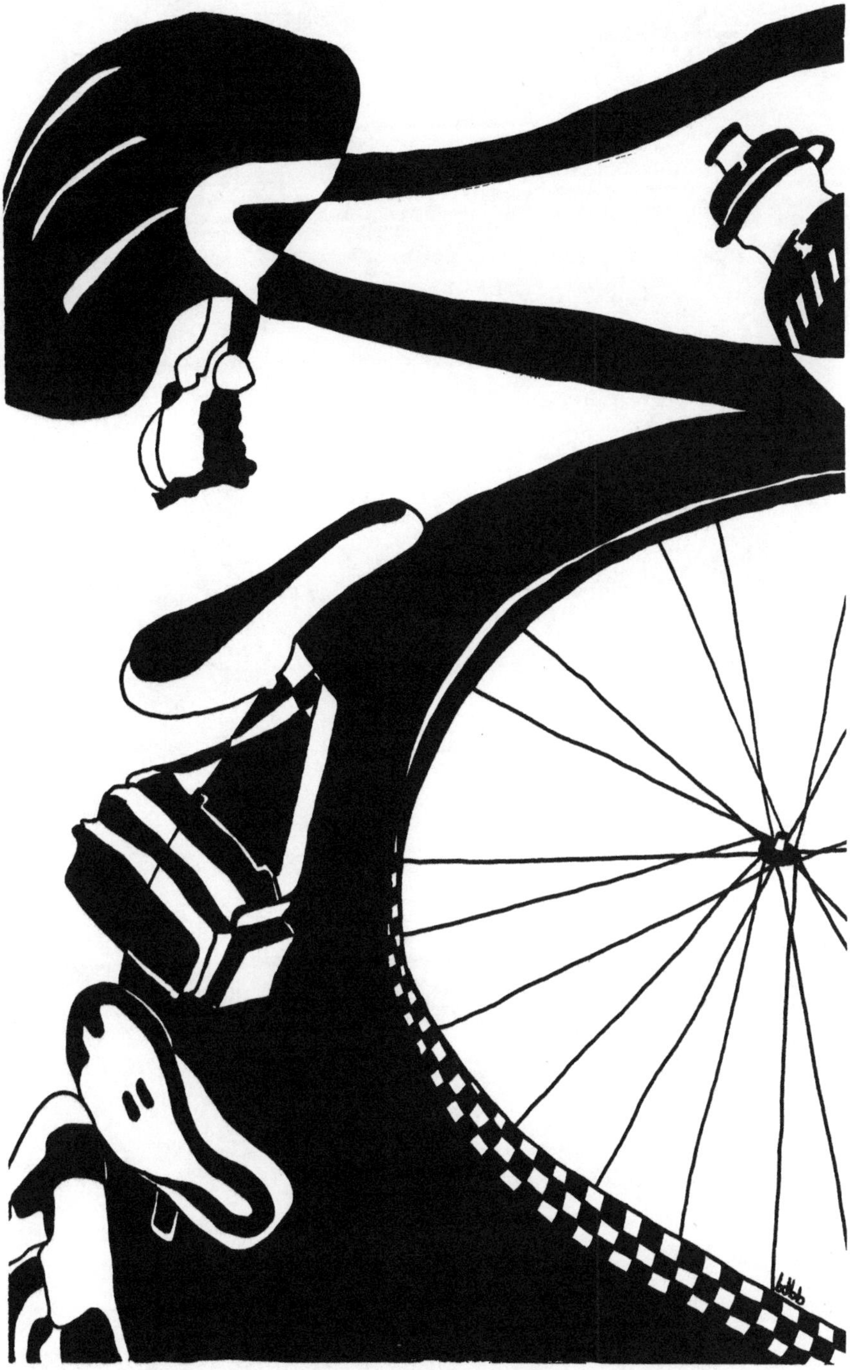

Chapter Four

A Post-Critical Aesthetics

The imagination required to understand a scientific theory, to appreciate a work of art, to respond to religious revelation, is a matter of degree. A scientific theory requires a lesser degree of imaginative effort to be understood. Once understood, our imaginative task has been achieved, and we may focus on the theory intuitively and effortlessly. The imagination is highly engaged throughout our time spent with an artwork. Art and religion require an exceptional thrust of our imagination that does not end: we never complete the activity of apprehending aesthetic and religious meaning.

The pursuits of making and beholding art, like science and religion, are cravings to make ever closer contact with reality. Each such craving demands its degree of imaginative power to be satisfied depending upon the difficulty of the task at hand. Discovery, be it scientific, artistic, or religious, requires greater thrusts of the imagination than does the apprehension of a theory. Yet, in each case, the craving to know is satisfied most fully when it grasps an idea which promises to reveal unfathomable implications. The promise of further revelation engages our intellectual passions which can never be fully satisfied. The idea grasped, contact with reality beckons the knower into ever deeper understanding. Dwelling in and breaking out of established perceptions of reality is pivotal to all pursuits of knowing.

However, our dwelling in a scientific, or even artistic, theory is not the same as the viewer's dwelling in a work of art. A theory indwelled becomes an extension of the self from which we pursue further contact with reality. A theory may be understood completely and applied imaginatively to subsequent apprehensions of reality, but a theory, once grasped, no longer requires imaginative effort to be understood.

An artifact, on the other hand, like reality itself, can never be fully understood. Our desire to understand an artwork elicits continuous thrusts of the imagination. The beholder dwells in the artifact, not with complete understanding, but with inquisitive passions into ever deeper revelations of its meaning. It is not a problem to be solved, but a dynamic event that reveals supra-natural reality as the viewer simultaneously dwells in the artifact and breaks out toward progressively deeper meaning. Art, to be experienced, commands more than a minute of our participation; the viewer must crave to understand it and attend to it for long segments of time. When s/he participates in the artwork, the beholder senses an excitement that urges onward; when meaning is grasped the imagination is further incited. The viewer becomes caught in an unresolvable tension between understanding the work of art and the artifact's continuous beckoning to unrealized meaning. The viewer, like the scientist, the artist, and the religious, craves to understand reality, the reality of the artwork. The scientist is able to attain complete (though tentative) understanding of a given discovery, and the artist experiences satisfaction when an artwork is complete. The viewer, however, is more akin to the religious who, in Polanyi's words, is caught in an

> heuristic upsurge which strives to break through the accepted framework of thought, guided by the intimations of discovery still beyond our horizon. Christian worship sustains, as it were, an eternal, never to be consummated hunch; a heuristic vision which is accepted for the sake of its unresolvable tension. It is like an obsession with a problem known to be insoluble,...[1]

Scientific discovery and artistic creation engage in problems which lead toward a solution; art appreciation and religion sustain our imaginative quest for extraordinary coherences of meaning.

A work of art arouses the viewer's passions to understand its event because it is an embodiment of extraordinary meaning which can be recognized as true. The visual arts are regarded as subjective and emotional. Science, on the other hand, masquerades as a detached, purely objective field of inquiry. The passionate aspect of scientific inquiry is disregarded. In truth, the emotive life, passion, is as much a component of science as it is art, religion, law, or any other field of knowledge.

Polanyi talks of the delight experienced by the scientist when discovery is attained. Scientific theories have an inarticulatable component that herald their beauty. This is essential to recognizing the truth of a theory.[2] Truth and beauty are expressed by both science and art. Through, what Polanyi calls our "intellectual passions," we rely on the disclosure of beauty to affirm a discovery.

Beauty in all cases of knowing evinces our contact with reality and it stimulates and attracts our passions. Intellectual passions, in turn, acclaim the appropriateness of beauty to its revelation of truth. The delight that guides inquiry and discovery's affirmation serves the scientist, artist, viewer, and worshiper. Beauty, by enlivening our intellectual passions, draws us toward discovery of an infinite and inexhaustible reality. We become fascinated and are charged with a wonder that leads us into an expanding interest in reality.

These mental passions are not our appetites that seek to fulfill purely private demands. Instead, intellectual passions seek to fulfill a public obligation to educate ourselves and others. They are a tacit response that recognizes the integration or disintegration of subsidiaries, and they thrust toward a new focal coherence. When the coherence is attained, beauty is recognized, the logical gap is crossed, discovery is realized, and public knowledge can be expanded.

To assert that art incarnates beauty is to state that art embodies a manifestation of the passionate spark that guides our contact with reality. It presents our tacit response to truth. When it does so, we affirm art's presentation as true. The reality presented in the artifact is supra-natural which is, in part, the subsidiary element that guides and interprets our response to natural reality. Art is passionate and communal. The artist does not create the artwork to satisfy her/his private need, but creates under obligation to culture and its demand that art guide its perception of meaning. As scientific passions guide our apprehension of natural reality, aesthetic passions guide our understanding of supra-natural reality.

Science is not a detached, purely objective enterprise any more than art is the result of subjective emotionalism. Both are passionate ventures that inquire into reality. They are feats of the imagination guided by intellectual passions. Science relies on beauty in its affirmation of truth; art embodies that beauty in its claim to represent truth.

If art represents facts of the imagination, how can it also be said to embody beauty? After all, the imagination is a tacit internal force of the mindbody, and beauty is the attractive element of reality outside of ourselves. How can art present both aspects? Sketches of the imagination, beauty, and intellectual passions can be rendered.

Imagination is the leader. Members of mindbody, such as intuition and passions, move through the imagination as they sort out clues of the unknown that point toward discovery and meaning. Imagination is imbued with a vision and implements movement on a path that leads, hopefully, to the fulfillment of that vision. The imagination does not discriminate; it will implement private appetitive demands as well as intellectual obligations to the public; it will execute simple performances as well as highly complex acts of integration. The imagination leads mindbody in its quest for truth.

Beauty is the lure. Beauty attracts, even harasses, the imagination by offering clues to reality. It is the unspecifiable element of reality that captivates the imagination. Beauty resides outside the self and draws the imagination into a deeper understanding of reality.

Intellectual passions are the guide that desire to know truth, recognize beauty, and share reality's discovery with others. As the imagination thrusts toward the unknown in its search for truth, intellectual passions provide a complex aggregate of clues to the imagination in order to apprehend beauty. Intellectual passions skillfully fuel the imagination. They are the connoisseurs of beauty. Intellectual passions dwell in beauty; the imagination breaks out toward truth. Intellectual passions are the interpreter and guide to the imagination which hammers out possible clues that will lead to reality's discovery.

The imagination implements the journey toward the promised land to which beauty beckons; intellectual passions incited by beauty guide the way. Finally, propelled by public obligation, intellectual passions embody truth's discovery in forms which may be shared. These forms include scientific theories, works of art, and religious expression.

Polanyi's triadic structure of knowing may be employed to summarize the inter-workings of beauty, imagination, and intellectual passions: beauty is the unspecifiable attracting aspect of a focally known reality; imagination is the seeker which aims for a meaningful relation of the subsidiaries to a focal whole; intellectual passions are the imagination's guide to discovery based on skills held subsidiarily. Intellectual passions rely on communal life to discover beauty; they are in turn affirmed by the community's recognition of beauty in the shared forms. Polanyi summarizes the idea as follows:

> A scientific theory which calls attention to its own beauty, and partly relies on it for claiming to represent empirical reality, is akin to a work of art which calls attention to its own beauty as a token of artistic reality. It is akin also to the mystical contemplation of nature....More generally, science, by virtue of its passionate note, finds its place among the great systems of utterances which try to evoke and impose correct modes of feeling. In teaching its own kinds of formal excellence, science functions like art, religion, morality, law and other constituents of culture.[3]

In answer to our question regarding art's ability to represent both facts of the imagination and beauty, we can now state that art represents facts of the imagination through an expression of beauty. Beauty demands to be given form so that it may be shared, be it through a scientific theory, a work of art, or a religious observance.

The artist submits to an unnameable, imageless reality but passionately seeks a tangible expression of beauty for others to grasp. Beauty relies on the independent existence of reality and therefore must be imaged by the artist in a way that is outside of the artist's self. The viewer is attracted by the expression of beauty in the artwork and desires to grasp its meaning. This is accomplished through feats of the imagination which are guided by intellectual passions that are able to recognize beauty and affirm art's presentation of truth. Art to be true must embody a beauty that can be affirmed, and it is the community's recognition of beauty that validates art's presentation of reality rather than illusion.

All domains of intellectual inquiry affirm their claim to truth, to knowledge, through passionate valuations which recognize some emotions as right. The rightness of these emotions guide the knower through the course of discovery. As the knower probes reality, intellectual passions function to attract or repulse the knower from clues that lead toward discovery. Intellectual passions enable the knower to distinguish between clues and facts that are prime consideration to a specific field and those that are not.

A skilled scientist is able to distinguish between a multitude of knowable facts which are of higher and lesser interest to science. A talented artist is able to distinguish which elements of form meet the commanding form's demands. A trained viewer is able to recognize an artistic classic. All are guided through an emotional response that can never be dispassionately defined. All depend on their tacit knowledge of beauty.

Our intellectual passions guide our vision of the general nature of things and thus enable us to interpret experience and its value. We are equipped to discern what is plausible, though all evidence may contradict it, and what is deceptive, though it is supported by evidence we cannot dismiss.

Polanyi characterizes three factors that contribute jointly to scientific value. These include (1) certainty, (2) systematic relevance, and (3) intrinsic interest. A deficiency in one can be compensated for by excellence in the others. Yet the deficiency in one is not cause to dismiss the factor all together. The joint application of these criteria protects science from valuing an irrelevant claim to knowledge based on strict objectivism, and from valuing subjectivism that has no bearing on reality due to its lack of basis in fact. Compensation is necessary in establishing scientific value; joint application of the factors is essential to avoid the perils of valuing one criterion without the stabilizing of the others.

1. Certainty pertains to accuracy of facts and their reproducibility. It is often the factor mistaken for the ideal of a strictly objective science. But because something is accurate, it is not imperative to science.[4]

2. Systematic relevance also tends toward the objective pole, since its recurrence can be predicted within a natural system. A unique event may have systematic relevance due to the profundity its discovery bears on reality. The discovery of a living coelacanth did not depend on the recurrent supply of such an animal for its scientific significance but on its existence as a common ancestor of all the land vertebrates. Systematic relevance is important for the breadth of its implications, even when no new maxims are established. It offers an authentic understanding of a large realm of experience. Systematic relevance then answers to the fittingness of the scientific facts within a given scientific framework. The coelacanth was a profound discovery within the scientific framework of evolution.[5]

3. While certainty and systematic relevance combine in determining the value of a scientific fact, intrinsic interest determines the value of subject matter. Polanyi points out that "living animals are more interesting than their dead bodies; a dog more interesting than a fly; a man more interesting than a dog."[6] A subject's intrinsic interest is rarely linked to accurate observation and systematic study. Usually, the more the intrinsic interest, the less dependable it is in observation. Thus, the compensation of one area's excellence for another's lack is essential to the establishment of scientific value. Physics is high in exactitude but low in intrinsic interest; biology is of great intrinsic interest but inexact.

These factors participate in the selective function of intellectual passions and affirm scientific value. They also inform aesthetic value. For though Polanyi characterized the three factors of certainty, systematic relevance, and intrinsic interest, as means to discern scientific value, his thought applies to art, and these criteria instruct aesthetic value.

1. Certainty and Iconography

Applied to art, the concept of certainty is a negative factor. Art is not imitation of reality, but a presentation of supra-natural reality. Art employs natural form, that may be interesting in itself, in order to point to a more profound reality that cannot be articulated. Therefore, the factor of certainty concerns itself with "windows to supra-natural reality." Does the artwork depict natural form so focally that it blocks our vision of the aesthetic realm or does its natural form function as a subsidiary vehicle to a more comprehensive whole?

Medical illustration is an art that is exact in its representation of natural form and does not beckon the viewer into a deeper inarticulate reality (Plate 3). Instead, it points the viewer toward a more articulate understanding of the actual form it depicts. What medical illustration presents is specifiable. The art of George Segal (1924-), on the other hand, is exact in its representation of natural form, but its form leaves a window to a deeper reality. The form functions subsidiarily as it points into itself to the aesthetic realm. When we focus on the exactitude of form captured by Segal, the meaning of the artwork is destroyed. The exacting form of medical illustration conveys its meaning as it points to the reality it portrays. Accuracy of observation makes medical illustration valuable to medical science as a tool for discovery; it is not discovery in itself. While Segal has chosen a form that is faithful to nature, this is not what makes his work valuable as art. The exacting form in Segal's work has meaning as it functions as a subsidiary element in the focal whole of the artifact which is its meaning.

Polanyi's epistemology characterizes the function of the tacit and focal elements within all structures of meaning. The tacit elements *have* a meaning that bears on something else that *is* their meaning. The elements that *have* a meaning deflect our attention away from themselves and in turn point our attention to that which *is* their meaning.

For example, a school photograph of a child *has* its meaning as it bears on the actual child who *is* its meaning. An artistic photograph of the same child *is* the meaning in which the live child's face functions subsidiarily and thus *has* meaning within the integration the artistic photograph achieves. The aesthetic meaning of the photograph is destroyed when the focus of attention is shifted away from the photograph to the child. The meaning of the school portrait is destroyed when the focus of attention is shifted away from the child to the photograph. The child and the photograph are not equal partners of an association but instead find their meaning in the from/to structure of knowing advanced by Polanyi.

Medical illustration *has* its meaning as it bears on the object which *is* its meaning. A medically exact drawing of the face *has* its meaning as it points to a live face which *is* its meaning. The exacting form of Segal's sculpture *has* its meaning in the artwork which *is* its meaning.

The particulars relied upon when achieving form need not be true to natural objects. Naturalism, which relies on objects in its search for form, and abstraction, which uses relationships, are components in all works of art. The naturalism and abstraction employed vary by degree in different works of art. George Segal's sculptures are highly naturalistic, Mark Rothko's (1903-1970) paintings are highly abstract. The precision of naturalism risks flattening the aesthetic dimension, abstraction endangers meaning if it does not echo some experience. Naturalism and abstraction, while always present in a given work of art, are employed in different mixtures. The natural-abstract mix avoids the perils of an art which is wholly illustrative or irrelevant to reality. All art lies between these two limits of naturalism and abstraction.

In art, certainty is a negative aesthetic factor insofar as naturalism must be mixed with abstraction to a greater or lesser degree. Certainty or an accurate rendering of an object is of value when it and the natural reality depicted participate subsidiarily in the artistic integration achieved. A degree of abstraction allows this to happen, and the form then serves as a window to a deeper reality. The form loses its value when its subsidiary function is deflected from the rendering to the natural reality portrayed.

This aspect of aesthetic value is a stylistic concern since it relates to how an object is represented. The degree of naturalism and abstraction within a work of art is one of the few stylistic elements that may be specified. However, even the categories of naturalism and abstraction can quickly elude analysis. Style is primarily an unspecifiable element in art. It integrates the particulars of the artifact into a harmonious coherence. Vague and profound, style is a permeating element within a work of art. It transcends observable artistic concerns such as iconography and form.

Iconography is the study of subject matter. It is an inquiry into what is depicted, not how it is rendered. It provides an index of elements such as era, geography, and race. An individual's clothes, the color of the skin, the story's motif, and the placement of individuals in relation to other individuals are all elements of subject matter that offer clues to an artwork's socio-historical context.

Iconography is not limited to representations of natural reality; it also includes the more abstract motifs, including color, line, shape, and texture. A description such as the painting contains a red field with patches of blue over the top third is an iconographic statement.

The simple observation of these facts of subject matter are of little value in and of themselves. For example, to state that Segal's *Walk–Don't Walk* (Plate 4) sculpture contains a Walk–Don't Walk sign and a curb with three people–two men and one woman–standing on it, is moderately significant. It may help us identify the sculpture and recognize what the artist is doing aesthetically. Or it may block our participation in the work of art's deeper meaning. Iconography cannot establish aesthetic value anymore than an observable fact in itself is valuable to science. It is when iconographic observations are joined with the factors of systematic relevance and intrinsic interest that they take on aesthetic value.

2. Systematic Relevance and Form

Systematic relevance in art considers the fittingness of an artifact to its more comprehensive style and the fittingness of formal elements to the whole of the artifact. Like iconography, systematic relevance concerns itself primarily with observable elements in works of art. It includes criteria that discriminate an artwork's originality from its continuance with tradition.

An artwork's style and form, however, are features of the artifact that we are aware of subsidiarily in their joint focal appearance. Style is a largely unspecifiable element that bears the imprint of the maker, the epoch, the race in the artwork. It is the cohesive element among a group of artifacts. Style is recognized through the art of connoisseurship. The beholder relies on a learned capability to judge and appreciate art acquired through training, tradition, and the art of seeing; an art connoisseur is able to recognize specifiable and unspecifiable elements that point to an artwork's style.

Form is more specifiable and thus more readily subject to analysis as it pertains to the formal elements that make up an artwork's composition. The formal features of visual art include line, rhythm, mass, space, light and shade, and color. These elements, when considered focally, are conceived as separate from subject matter. A line is judged for its thickness, direction, continuance, and materiality rather than as the edge of a face. The representational quality, if any, is not of focal concern when a line is pondered formally. This is similar to our ability to think of the abstract number seven without thinking of seven specific objects.[7]

A formal analysis of Segal's *Walk—Don't Walk* could include the following observations: the sculpture relies on a strong vertical axis that is dominated by the Walk—Don't Walk sign, and repeated by the three figures. The sign and two of the figures present straight upright verticals, while the figure closest to the sign breaks the static rhythm with a more fluid relaxed vertical. The vertical lines are supported and framed by a geometric horizontal base. This shape is repeated on a vertical axis by the square sign that tops the vertical sign pole. The tops of the three heads are points on a horizontal triangle that give the sculpture depth and unity. These are a few of the formal elements that may be observed. The set of complex relations that these formal elements present are appreciated for the design they embody.

These observable relationships that jointly embody the form are not primarily, if at all, indicative of tangible objects. The essential relational quality between the formal elements of art, be they highly naturalistic or abstract, incarnates a sensuous presence that does not rely on denotation. The design is striking in and of itself.

A Rothko painting may be described in terms of line, shape, and color as readily as the Segal sculpture. The visual arts through articulation of formal elements create visual relationships that invite us to understand and enjoy them. However, when we examine the formal elements as interesting in themselves, the comprehensive meaning of the artwork is dismembered. We need to grasp form while imaginatively attending to the joint meaning on which they bear.

Art criticism disattends from the imaginative integration of the extensive whole. Analysis of form or style is destructive to the comprehension of the artwork, at least while the analysis is taking place. When we attend to the particulars focally, they cannot function subsidiarily; their integration is destroyed. While analysis destroys the overall quality, a deeper understanding of the artwork is gained through subsequent integrations. Enriched particulars are carried into their new subsidiary participation.

We must enter analysis with care realizing that it is not the end-all to understanding art. The analysis inherent to art criticism enlarges our experience of art *if and when* we return to the comprehensive whole of the artifact. Too many art historians rely on their analysis of the particulars so heavily that they have obscured beyond recall the comprehensive meaning of the artwork.

An artwork is relevant to the history of art due to its profundity. It offers a new creation that bears on our understanding of supra-natural reality. Its relevance may be a deepening continuance of existing formal and stylistic discoveries. Such is the work of Claude Monet (1840-1926). His paintings were new creations that relied on, yet enriched, the Impressionistic immediate style and use of fleeting form.

Paul Cézanne (1839-1906), on the other hand, dwelled in Impressionism but broke out of its frame, presenting a novelty that altered the course of art. Cézanne moved beyond the Impressionist's visual reflections of light to the essence of solidity. He incorporated the firm structure of the old masters, especially Nicolas Poussin (1594-1665), with the chromatic sequences of Impressionism. The result was a new orientation to painting that integrated geometrical organization with color harmony. The profundity of Cézanne's work lies in its unique contribution that continues with tradition while showing the way to new developments, especially Cubism. The relevance of Cézanne does not rely primarily on a stylistic and formal continuity as much as it does on its novelty. Cézanne broke open style and form in a way that led to a changed understanding of supra-natural reality.

Form, like iconography, relies on elements that may be observed within a work of art. These elements may even be systematically articulated. The value of art, however, does not fully rely on articulation of formal or iconographic elements. Descriptions of them are useful, but they do not determine art's value. To base art's value on its specifiable elements would reduce an artwork's meaning to those elements. The integrated whole in which the form and iconography participate subsidiarily cannot be fully articulated.

3. Intrinsic Interest

The third factor of aesthetic value, corresponding to Polanyi's description of intrinsic interest, and working jointly with iconography and form, enables us to more deeply understand the inarticulate aspects inherent to works of art. Intrinsic interest describes less specifiable aspects of an artifact than the observable elements of iconography and form. Intrinsic interest joined with iconography and form points to the value of an artwork. It does not rely on iconography and form. In fact, in most cases the more formalized the artifact, the less its intrinsic interest. Recalling our definition of reality as what reveals itself indeterminately in the future, you could argue that abstract expressionism is more real and thus more interesting than naturalism. The use of an object as a guide for form can be helpful in the process of expression and appreciation, but it can also be binding. It can be more difficult to deepen our experience with a naturalistic painting than a non-representational one.

Abstract art is initially more difficult to indwell and can appear uninteresting. (I am using the term "abstract" in reference to non-representational art. "Abstract" is a relative term, since all art is more or less abstracted from natural reality.) What is interesting about three black lines on a white field? Or a solid black canvas? The difficulty is due to non-representation's immediate demands on the viewer's imagination. The beholder may not approach the artifact with detached observation and assume that its meaning is being grasped. The screen of observation must be dissolved at once as the viewer pours his/her self into the artwork. And abstract form seems fecund with meaning once the viewer submits to it. For example, the solid black field of an Ad Reinhardt (1913-1967) painting reveals a canvas that dances with nuances of geometric shapes. Naturalism's form, on the other hand, is easily recognized and thus less demanding of the imagination initially; its natural form can block us from penetrating its depths. Andy F. Sanders states the idea:

> We may expect that the more an entity is real..., the more intrinsic value it possesses. Since there is also a correlation between levels and degrees of tangibility, it seems we are also allowed to say that the more intangible a comprehensive entity is, the more intrinsic interest it possesses. For instance, 'intangibles' like problems and the human mind are said to be more real and more intrinsically valuable than the directly observable natural objects so cherished by empiricism. In other words, the lower the level of being, the more easily comprehensible an entity (say, a pebble), the lower the degrees of indwelling and intrinsic value, and vice versa.[8]

Non-representational art is less easily grasped than naturalism. Thus, abstract art possesses more intrinsic interest than naturalism.

The compensation of excellent iconography or form for the lack of intrinsic interest is essential to the establishment of aesthetic value. Naturalism is high in exactitude but low in intrinsic interest; non-representational art is of great intrinsic interest but inexact. The synthesis of form, iconography, and intrinsic interest protects aesthetics from valuing one style of art over another. One criterion is stabilized by the others. When we dwell in an artifact of aesthetic value, the artwork will prove to be rich in content and polyvalent regardless of its relation to natural form. Art cannot be reduced to a single statement or nuance. Its whole is more than its parts, and the content takes shape through the mix of aesthetic criteria.

Unfortunately, Polanyi misses this important feature in his discussion of art. He draws a line that separates representational art from abstract art. He describes representational art as a form that makes an affirmation which can be expressed verbally. Visionary (abstract) art is different because it "says nothing that can be expressed in a statement."[9] Polanyi is in a double bind. When considering naturalism he disregards art's essential feature of making the tacit visible, and he misses subsidiary nuances of non-representational art that can be expressed verbally. This problem is also common to Paul Tillich's aesthetics; it arises from trying to discuss a subject without dwelling in it openly and fully.

The aesthetic value of an artwork relies on the joint application of style, iconography, form, and intrinsic interest. These criteria work together in different balances, just as the criteria of certainty, systematic relevance, and intrinsic interest function jointly in judgments of scientific value. I have described (1) art's demand of immediate indwelling as pertinent to intrinsic interest. But art's intrinsic interest refers to more than this demand. Two other features of intrinsic interest include (2) art's metaphoric quality and (3) art's value to a specific field of inquiry.

Polanyi describes the role of metaphor in his theory of semantic meaning. His theory includes how indication, symbol, and metaphor relate in our pursuit to make meaning.

Semantic meaning refers to all forms of meaning contrived by humanity as distinct from the meaning of perception, skills, and other part-whole relations found in nature.[10] Indication, the first type of semantic relations, can most easily be understood in relation to the from/to structure of language. A word *has* its meaning in the message which *is* its meaning. The word's meaning can be destroyed if we focus on the word and the sound it makes on our lips. A word repeated for no reason other than to make its sound quickly loses its meaning. Words function as pointers to their focal meaning. Words have no intrinsic interest in themselves, but point to something that is interesting.

Medical illustration serves subsidiarily as an indication: it is of little interest in itself but points to an object of interest. Likewise, a denotative photograph of an artwork is indicative. Artistic photography follows all that I assert regarding the visual arts. A photograph of a person is meaningless without knowledge of the subject, and a photograph of an artwork is meaningless apart from the actual artwork. An artifact cannot be appreciated through a photograph anymore than a person can be understood through a picture.

When we invert this process so that "matters of great intrinsic interest take up the subsidiary position, while *from* these we attend focally to things having no intrinsic interest,"[11] a symbol is created. A flag has no intrinsic interest; it is a meaningless piece of cloth. But when we attend to our nation's flag, we do so by attending subsidiarily from our membership in the nation and the memories learned and experienced in that nation. We embody our life in the nation in the flag. The meaningless piece of cloth attains meaning from our subsidiary integration of our understanding of the nation to that cloth. The piece of cloth stitched into a flag *stands for* the nation; the cloth symbolizes the nation. Designating a nation by its name integrates the name to a country. In an opposite way, symbolizing that nation by a flag integrates a country to a flag.[12] In sum, a word, map, photograph is intrinsically uninteresting but indicates something that is of focal interest; a symbol, the object of focal interest is intrinsically uninteresting but stands for something of intrinsic interest that is held subsidiarily.

The operative difference between indication and symbolization is the relation of the self to the object of focal interest. The self integrates the clues of indication into coherences that are projected away from the self. A word is understood from clues dwelled in by the self to the word's meaning. "The self is never carried away in indication; it is never surrendered or given to the focal object...indications are always *self-centered*. By contrast, symbolizations are *self-giving*."[13] More than an integration of clues directed *from* the self *to* the focal object occurs in the grasping of a symbol's meaning. An added process takes place in which the self carries diffuse memories and experiences *into* the symbol. The symbol becomes a visible embodiment of a comprehensive whole that includes both the integration of the symbol to its meaning and the integration of the self into that meaning. The *from/to* structure of knowing is enlarged to include an embodiment of the self *into* the symbol's meaning. The self is surrendered into the symbol as it is carried away in a self-giving act of integration.

I agree with C. Phil Mullins[14] who finds curious that Polanyi classes indication in the family of semantic relations that includes all human-constructed cases of meaning as distinct from the meaning we meet in nature. Indication follows the same structure of from/to knowing as skills, perception, and other part-whole relations. Self-giving is the added element that expands our from/to knowing to include "into knowing." Self-giving is a highly imaginative complex integration that requires self-surrender. Polanyi would be more consistent to include indication with natural meaning and to originate the distinction of semantic relations with symbol. Semantic relations would then be continuous but distinct from natural integrations due to their ability to carry the knower away. Since occurrences in nature, such as a breathtaking sunset, can also carry the knower away, perhaps the distinction should be between natural and supra-natural integrations.

That art consists of from/to and into integrations is my understanding and extension of Polanyi's thought. This idea enhances epistemological concerns regarding the discernment of an artwork's meaning. "Into" is at times grammatically awkward. To help ease this grammatical tension, the structure of knowing art can be spelled out by stating that the viewer attends to and surrenders into the artifact.

Adding surrender (Polanyi's idea) poses its own problem, for it is suggestive of mysticism and oppression. Neither of these are intended. Through the use of the terms "into" and "surrender into," the high degree of mindbodily participation is suggested. The self is carried inside the work of art and made a part of its form. Appreciating a work of art is like looking into a mirror, but more than the self is reflected. The self is reflected as an integrated component of all that the artwork presents.

Surrender does not involve the giving up of oneself; when the self is given into the artifact, the self is given back to the viewer with a clarified, enlarged understanding of both the artifact and the self. Surrender does suggest that control of the situation is relinquished so that novelty may be grasped. When a viewer attempts to manage his/her experience with an artifact, be it through art-critical analysis or through a pre-conceived judgment, the artifact's meaning will be unattainable. Judgments, analysis, and socialization are surrendered into the artifact. They must not be used to restrict the artwork's meaning.

Surrender overcomes subjectivism; it involves a recognition and seeking of art's intrinsic intelligibility. The beholder submits mindbodily to art's reality. Finally, surrender demands that the viewer approach the artifact with heuristic passion; the viewer needs to be open to being changed by the experience with the work of art.

Metaphor, too, is self-giving. This third type of semantic relation is intrinsically interesting both in itself and in its subsidiary matters. "When a symbol embodying a significant matter has a significance of its own and this is akin to the matter that it embodies, the result is a metaphor."[15] The exacting naturalism of Segal's sculpture is interesting in itself as is the comprehensive form of the artwork. But if we focus on the fidelity to nature rather than the comprehensive form in which it participates, the metaphoric meaning is destroyed. While the subsidiaries are of high intrinsic interest, the meaning of a metaphor can only be achieved through its integration. The parts may be meaningful in themselves, but their meaning participates in the focal whole to achieve a novel meaning.

The comprehensive meaning of a metaphor results from the integration of the subsidiary elements to the focal object plus a further integration of the self into the united subsidiaries and focal object. The focal object of the metaphor functions focally and subsidiarily. The integrative process of a metaphor can be described in three phases. First, as in indication, the subsidiary elements of the metaphor–the tenor–bear on the focal object–the vehicle. Second, like symbolization, the vehicle bears back on the tenor, thereby embodying the tenor in the vehicle. Third, the viewer's self is integrated into the united tenor and vehicle so that the self is moved deeply and is carried away and possessed in the metaphor (Figure 1). This is an imaginative process in which two or more disparate matters are integrated into a novel meaning that cannot be expressed in another way.

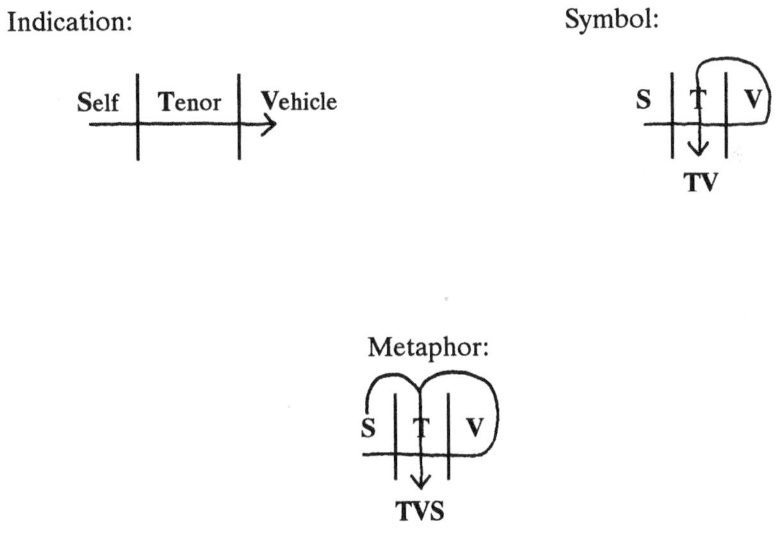

Figure 1. The Workings of Metaphor

While indication functions subsidiarily and *has* meaning as it points to the object which *is* its meaning, symbol functions focally and *has* meaning as it embodies the subsidiaries which *are* its meaning. And metaphor functions focally and subsidiarily and *is* meaningful while it is integrated into a whole which *is* its more powerful meaning. Indication follows the from/to structure of all knowing; it is self-centered. Symbol and metaphor include an additional self-giving process that embodies the self into its meaning.

The translation of the metaphor into prose is destructive of its meaning and more often than not results in nonsense. Segal's *Walk–Don't Walk* verges on the absurd when we state that three life-sized bronzed figures stand on a curb waiting for the "Walk–Don't Walk" light to change. The figures cannot possibly walk since they are not alive, so how can we assert that they are "waiting"? Bronzed figures don't "wait"; they are just lifeless shapes that resemble human form. Polanyi, in referring to works of art, explains how metaphor moves beyond absurdity to a deeper more unitive meaning, "but their metaphorical expression is clear and forceful. For it makes no absurd claims for an absurd eventuality, but affirms these [elements of metaphor], because it means by them something else than what they mean in themselves."[16]

How has Segal, in metaphor, achieved a powerful and novel meaning? He integrates disparate elements that do not occur naturally into a unified meaning. The bronzed figures are incompatible with the actual "Walk–Don't Walk" sign: the sign is a part of our everyday experience, but the figures are not. Segal invites us to integrate these incompatible elements in order to discover the sculpture's comprehensive meaning. To do so may be difficult; it demands we cross the same logical gap Segal had to cross to create his sculpture.

The metaphoric quality of an artwork necessitates direct experience with the artifact. A photograph of an artwork is indicative and cannot carry us into the artwork's power. An immediate proximity to the artifact enables us to surrender ourselves into it and grasp its meaning. A photograph is a tool that may aid analysis, but it cannot hold the comprehensive entity that moves us ever more deeply into its meaning. Tillich's experience of art during World War I is exemplary. While viewing reproductions of art helped to relieve Tillich from the terrors of the war, not until he had a direct encounter with Sandro Botticelli's *Madonna and Child with Singing Angels* (Plate 5) did he experience the revelatory power of the artwork. As he has stated:

> That moment has affected my whole life...the experience goes beyond the way we encounter reality in our daily lives. It opens depths experienced in no other way.[17]

The third feature of an artwork's intrinsic interest is its value to a specific field of inquiry. A sculpture by Michelangelo is of more interest to a Renaissance historian than it is to a scholar on Mid-Eastern art. An artifact with discernible theological dimensions is of more interest to a theologian than a creation that does not present a theological meaning. This is the selective function of intellectual passions which affirms the intrinsic interest of an artwork in relation to a field of inquiry. When the viewer dwells in the artwork, s/he appreciates the artifact through an intellectual framework that seeks clues of interest to their field. A theologian who is a connoisseur of art is able to distinguish the content of an artwork that is of interest to theology.

Yet more than affirmation of meaning is required of the viewer. We must be open to the artwork and its demand that we change our view of reality and subsequently our intellectual frames into a new and deeper vision. Our mindbodily being is immersed and transformed as we attend to the supra-natural reality embodied in the artwork. Polanyi designates this as the heuristic function of intellectual passions. We do not merely affirm art's reality, but evoke clues which will lead to an indeterminate future discovery. Passion guides and sustains our strivings to follow these intimations to deeper understandings, and as a result, we see differently.

Creation involves the artist's persistent pursuit of clues that point toward an as yet-unknown painting. The artist, like the creative scientist, relies on heuristic passions to solve a problem. The artist recognizes a problem of an indeterminate form and sets out to discover an embodiment of an indeterminate meaning. The work is creative because it changes our view of the world by deepening our understanding of it. The change is irrevocable, we are made to see and think differently.[18]

When a scientist is posed with a problem, s/he relies on a known set of clues that point toward an indeterminate, unknown focus. An inventor's problem is the opposite. The inventor works toward a fixed focus and seeks the subsidiary elements that will achieve the inventor's aim. Invention is like the accomplishment of a skill: the aim of the skill is known, while its means of performance is largely unspecifiable. The inventor's task is to discover specifiable elements that will make a determined technological performance possible.

The artist's task is indeterminate at both ends: the aim of embodying meaning and the means of its achievement are indeterminate. The artist "searches for means of solving a problem conceived for this very purpose, and pursues this quest while continuing to shape this problem to fit the means for solving it."[19] When the artist works toward creating the artifact, the artwork takes on a life of its own. The elements of concretization resist and invite form. The commanding form responds to the form causing a change and shift in its aim. The process of creation continues to alternate between the quest for form and the consequential clues to its meaning, adjusting in light of the emerging form. The artist continues to seek the shape of the as yet-unknown artifact while simultaneously seeking its meaning. As s/he does so, s/he crosses a heuristic gap that lies between the problem of an as yet-unknown painting and its discovery. Once the discovery is achieved, the artist will no longer be puzzled by the problem. The artist will become a different person who sees and thinks differently. Such is the heuristic function of artistic discovery.

The beholder of art, too, is guided by heuristic passions. The viewer, like the artist, is posed with an indeterminate set of clues and focus. The concrete elements in the artwork are known elements that point toward an indeterminate meaning. Yet, the artifact is a fixed focal entity that reveals subsidiaries that are unspecifiable. The act of art appreciation echoes both the task of the scientist and the inventor. The viewer seeks the artifact's meaning as unspecifiable subsidiaries are simultaneously discovered. The beholder is posed with the problem of interpreting an unknown artwork and must cross a heuristic gap to discover its embodied meaning. In doing so, the viewer's interpretive framework is changed. That change is irreversible. The viewer's vision is modified, and s/he shall never see the artwork—nor the world—as before. The viewer experiences delight as s/he acclaims the artwork's beauty while recognizing the truth embodied. The beholder, like the artist, is guided, sustained, and affirmed by the imagination and intellectual passions.

The viewer approaches the artwork from a specific interpretive framework, but does not discover the artifact's meaning solely by applying that framework to the artwork. The beholder's location or life experience is an indispensable clue to the understanding of the artwork, but it does not determine its understanding. The viewer commits to the artifact's intrinsic intelligibility with a belief that it presents the truth. In doing so the viewer is impelled to make a heuristic leap across a logical gap and dwell in the new framework called forth by the artifact. The viewer does not passively observe the artwork; s/he actively dwells in it.

The viewer is called to surrender the self into the work of art, to attend to its imaginative integration with great interest without analyzing it. The beholder resides in the visual presentation following the heuristic upsurge which strives to break through accepted frameworks of thought. S/he is guided by intimations of understanding supra-natural reality and thereby apprehends the artwork's meaning. When meaning is grasped, subsidiary elements come to the fore. When the viewer subsequently surrenders into the artifact, its meaning is enlarged to consequently reveal new subsidiaries *ad infinitum*.

Doug Adams has developed several methods to aid the viewer in surrendering the self into the artwork to facilitate indwelling so that an artwork's rich potential of meaning may be discovered. The methods include having conversations with the piece, asking of the art, for example: What is my name? How loud am I? Am I warm or cold? What kind of energy do I have? Adams proposes other methods for a kinesthetic response to the artwork. Since all knowledge is mindbodily rooted, how much better to get the body actively involved from the start. Adams suggests that we take the same pose of a given figure. In the absence of a figure, he suggests we choose a color, assume a shape of that color, make a sound that expresses that color, and finally a movement. Or we may bodily move through the painting by walking, skipping, dancing the lines and shapes. All of these methods suggest ways for us to indwell by deploying the body in tacit ways so as to attend focally to the artwork's meaning.[20]

The theologian may ask the artwork questions that lead to theological meaning, for example: What traditional or cultural iconography is used? Does the artifact tell a story? Is the story familiar, new, open-ended? Does the work of art have a religious style which is prophetic (asymmetrical, rough-edged, disorderly) or priestly (symmetrical, harmonious, orderly)? Or, does the artwork re-create the ordinary into a value, or enhanced experience? Does it make me see the obvious in a new and different way? Does it open the truth within the ordinary?

These questions do not suggest that the beholder fabricate a subjective meaning, but they facilitate indwelling in which the viewer submits to and into the power and integrity of the artifact while simultaneously dwelling in her/his unique store of tacit knowledge. The viewer consciously indwells by surrendering himself/herself into the experience of the artwork's greatness.

If, however, the viewer attends to the painting critically, the focus of attention shifts away from the painting to critical concerns. The awareness of the painting then becomes subsidiary, and its aesthetic power is lost. We have established that the loss of an artifact's power can be due to shifting our focal attention away from the comprehensive whole to its parts or to other focal concerns. Tacit knowing provides an explanation of our ability to interpret the same set of particulars in different ways. The phenomenon that explains how a group of artists working from the same model each depicts the model uniquely, also governs the principles of art appreciation.

A painting can be viewed from different frameworks which may be simply alternative ways of viewing the painting in question or may be progressively more penetrating when a heightened imaginative effort guides skilled attentiveness and insightfulness. Thus, one viewer may merely see the portrait of a woman when viewing Cézanne's portrait of his mother, *Madame Cézanne in a Red Dress* (Plate 6), but a connoisseur of art may see color theories employed, the asymmetry of the figure, one-half of the face being much more mature than the other,[21] the relationship of this painting to other paintings by Cézanne, and by other artists. Two trained art critics may approach the painting differently due to their intellectual horizons, including cultural heritage, age, and training. Their ensuing experience of the painting will differ from one another. Gradually, within the larger framework of their field or the community of interpreters, one view will be accepted as more adequately integrating the particulars of the painting according to the community's criteria—or perhaps the community will integrate both views so as to deepen the experience of the painting more than either view could individually.

An individual not trained in art criticism yet being a lover of art may cultivate art connoisseurship through hours of engaging with artworks. S/he will be able to grasp a quality of the painting unavailable to someone who has not cultivated an openness to seeing and experiencing art. An inability to grasp art's meaning may be due to lack of interest or exposure but it also may be due to our inability to focus on the joint meaning of the particulars because we are enslaved by a specific intellectual framework. In this case the beholder insists on focusing on particulars, concerns, or ideologies instead of the comprehensive artifact.

In sum, the factor of intrinsic interest contributes to aesthetic value in at least three ways: the immediate demand of indwelling, the metaphoric quality, and the value to a specific field of inquiry. All of these function jointly toward our discernment of an artwork's content and its intrinsic interest.

4. Aesthetic Merit and Content

Style suffuses all of the factors (iconography, form, and intrinsic interest) that contribute toward an appreciation of aesthetic value with its own distinctive quality. Style is a largely unspecifiable element innate to all works of art. Within each of the factors that contribute to aesthetic value clues can be identified that point to style. How a figure is dressed, the quality of line, the degree of abstraction employed, are a few examples of elements that point toward style. Since style permeates the criteria of iconography, form, and intrinsic interest, it may be recognized as the glue which holds the criteria of aesthetic value together. Yet the work of art's comprehensive meaning is more than its style. While style unifies the artifact, it does not determine its meaning.

Thus, style, iconography, form, and intrinsic interest participate jointly in the establishment of aesthetic value, but they do not circumscribe the artwork's meaning or content. As Polanyi states, "we are aware of them focally in terms of what they add to that meaning and affect its quality."[22] The subsidiaries, rich in intrinsic interest, imbue an artwork with a focal content that cannot be explicitly specified. The criteria of aesthetic value describes the issue of compensation at work within different works of art. The factors are inherent to all art, but they participate in varying combinations and proportions.

In 1962, Polanyi worked out three further criteria which guide the scientific community in its recognition of the merit of a scientific contribution.[23] Like the factors of scientific value, they function together in a mutual relationship, and help us understand aesthetic merit and content. Content is the joint meaning of the comprehensive artifact. It is that element of art that cannot be wholly articulated, but only experienced. The criteria of aesthetic merit function together to describe partially specifiable elements of an artwork's content.

The first scientific criterion is plausibility. The contribution must be fitting to "the current scientific opinion about the nature of things."[24] Art, too, must be plausible in the sense that the community of interpreters recognizes its beauty and is able to affirm its presentation of truth. If the work of art has not beauty, if it rings false or is wholly absurd, it will lose its aesthetic merit.

The second criterion is Polanyi's description of the three factors of scientific value, certainty, systematic relevance, and intrinsic interest. Aesthetic value includes style, iconography, form, and intrinsic interest.

The third criterion of scientific merit speaks to the originality of the scientific contribution. Polanyi measures originality by the degree of surprise caused by the contribution to its community. He explains:

> The unexpectedness of a discovery will overlap with its systematic importance, yet the surprise caused by a discovery, which causes us to admire its daring and ingenuity, is something different from this. It pertains to the act of producing the discovery. [25]

In the case of art, originality moves beyond current conceptions to a presentation of a reality yet-unknown. It may or may not have systematic importance, but it is always rooted in the artist's tacit location. This criterion brings the making of art into a mutual relationship with its appreciation. The appreciation of art's beauty and its aesthetic value function jointly with art's making when aesthetic merit is being assessed. Beauty and aesthetic value are indwelled and enable the artist and viewer to break out into recognizing originality and grasping its novel apprehensions of reality.

The artist creates a new comprehensive whole which satisfies the standards of appreciation that the artist sets for herself/himself. The standards of aesthetic merit specified, in part above, must be satisfied, and these standards are set by the artist's intellectual passions. Once completed, the artwork demands recognition by its viewers, who in turn, dwell in the artwork by the same self-set standards of aesthetic merit. Persons maintain these standards of values that guide and judge our achievements. When beauty, aesthetic value, and originality are recognized, the meaningful content of an artwork arises.

NOTES

1. Michael Polanyi, *PK*, p. 199.

2. Ibid., p. 133.

3. Ibid.

4. Ibid., p. 136. See also Michael Polanyi, "The Republic of Science," in which Polanyi speaks of accuracy (p. 54); and "The Growth of Science in Society," where he talks about exactitude (p. 83), both in *KB*.

5. Polanyi, *PK*, p. 137.

6. Ibid., p. 138.

7. See Jane Dillenberger, *Style and Content in Christian Art* (New York: Crossroad, 1986), p. 19.

8. Andy F. Sanders, *Michael Polanyi's Post-Critical Epistemology: A Reconstruction of Some Aspects of "Tacit Knowing"* (Amsterdam: Rodopi, 1988), p. 235.

9. Michael Polanyi, "Visionary Art," in M:P, p. 7.

10. Polanyi, "From Perception to Metaphor," in M:P, p. 18.

11. Ibid., p. 17.

12. Ibid., pp. 17-18.

13. Michael Polanyi and Harry Prosch, *Meaning*, p. 74.

14. C. Phil Mullins, "Polanyi's Analysis of Art: A Critical Appraisal," a paper delivered at the Centennial Conference on the Thought of Michael Polanyi, Graduate Theological Union, Berkeley, Cal., 15-16 March 1991, pp. 7-9.

15. Polanyi, "From Perception to Metaphor," in M:P, p. 23.

16. Ibid., p. 23.

17. Paul Tillich, "One Moment of Beauty" (1955), in *On Art and Architecture (OAA)*, John and Jane Dillenberger, eds. (New York: Crossroad Publishing Co., 1987), p. 235.

18. Polanyi, *PK*, p. 143.

19. Polanyi, "Works of Art," in M:P, p. 23.

20. Doug Adams and C. Phil Mullins, "Meaning with the Arts," pp. 33-5.

21. See Doug Adams, "Art Affirms Times to Mourn and to Dance," *Church Teachers* (September-October 1988), pp. 58-61.

22. Polanyi, "From Perception to Metaphor," in M:P, p. 27.

23. See Polanyi, "The Republic of Science," in *KB*, pp. 53-54.

24. Ibid., p. 53.

25. Ibid., p. 54.

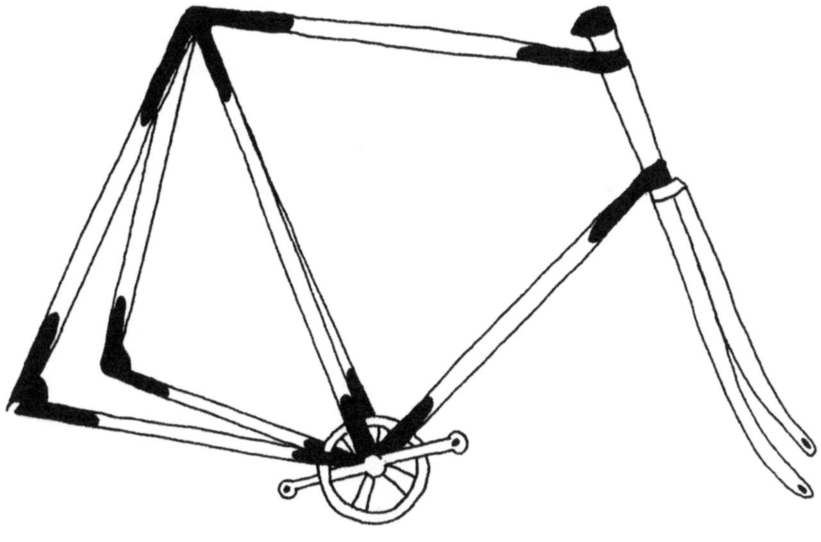

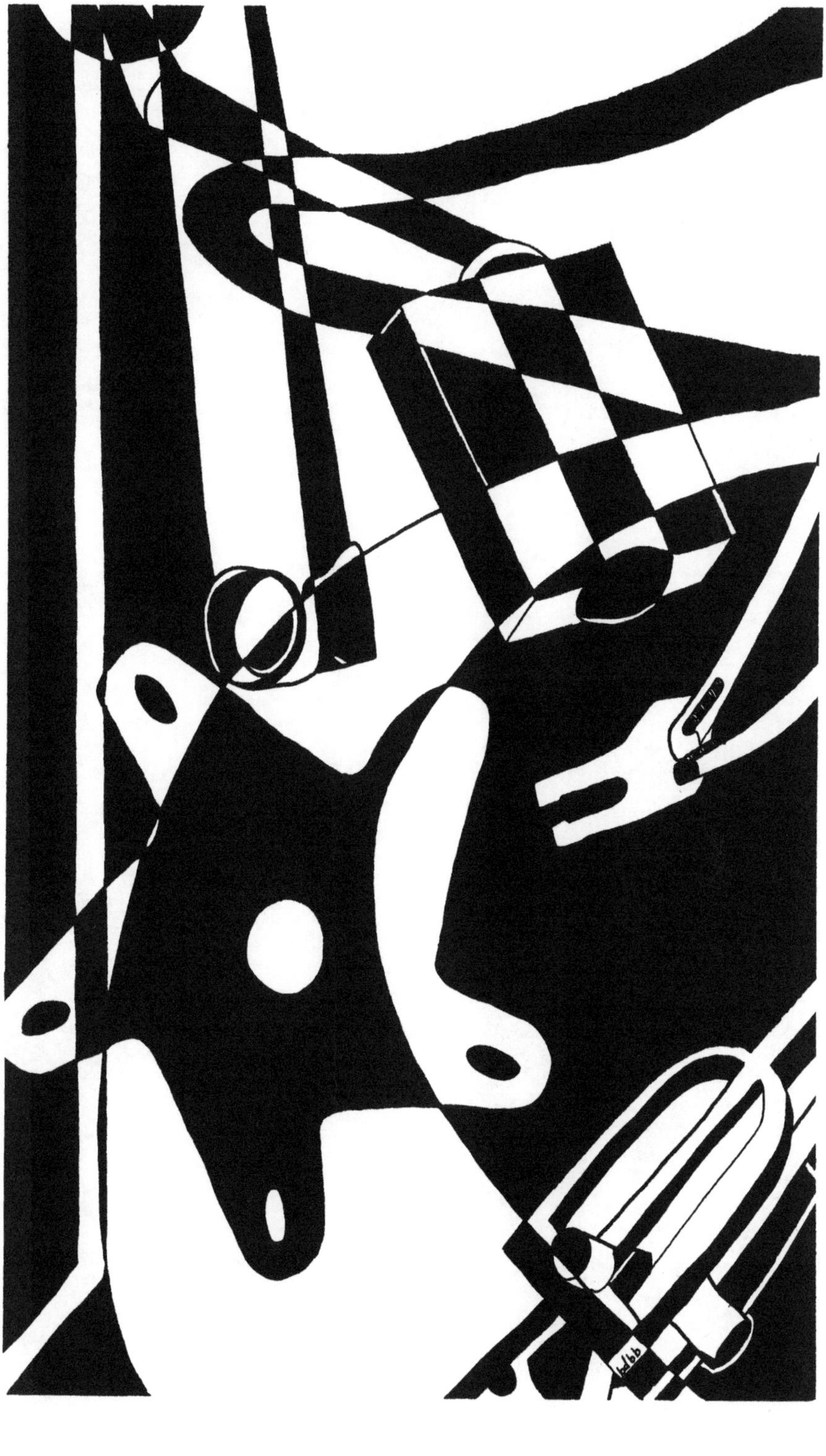

Chapter Five

The Structure of a Stratified Universe

Polanyi describes our world as a stratified universe in which "the upper level relies for its operations on the laws governing the elements of the lower one in themselves, but these operations of it are not explicable by the laws of the lower level."[1] The uppermost level of a thing is its most comprehensive feature, and it cannot be specified in terms of the lower level.

We have seen that rules are helpful in learning a skill but that they do not explicate the practice of an art. Rules are a set of particulars that are integrated into the higher level of a skillful performance. A skill is a comprehensive performance that draws upon and governs a complex integration of maxims, motoric action, intention, training, and other unspecifiable elements of our mindbodies.

The instance of a clock provides an example. If a person takes a clock apart and examines each part carefully, s/he will be unable to determine the principles that govern the clock's ability to keep time. The purpose the clock serves and how it works will elude such an investigation. The parts of the clock are matters of physics and chemistry which constitute the lower level of the clock. Only through knowledge of the upper level of the clock, the level that describes the clock's operational principles, can the clock's purpose and means of keeping time be detected. The upper level is the most comprehensive feature of the clock, and it is unspecifiable by its parts. The upper level relies on the lower level of physics and chemistry for the establishment of its material conditions and its success and failure. An analysis of the parts can determine why the clock is not functioning properly; the integration of the comprehensive entity describes the successful working of the clock.

Between the two levels a logical relation exists. The operational principles of the clock are both made possible and limited by the lower level of physics and chemistry, though the upper level is not determined by the lower level. The laws of physics and chemistry determine the conditions for the successful operation of the clock, but within these laws an area of possible variation is left open. The system of laws is open rather than closed. The operational principles of the clock harness the areas left open in order to establish a purpose for the physical parts that would not exist otherwise. The operational principles of the clock impose "boundary conditions" on the laws of the lower level. Thus, the logical relation between the two levels is of "dual control." The clock functions because the parts have specific properties *and* because the parts are integrated and controlled by principles that cause them to work together. Both levels are essential to the working of the clock, but their status is not symmetrical. The lower level is loses meaning apart from the upper level which defines that meaning; the upper level relies on the properties of the lower level to achieve that meaning.

The logical relation of the two levels corresponds to the two terms of tacit knowing, the subsidiary and the focal. The upper level corresponds to the functional aspect of tacit knowing in which we rely on our awareness of the particulars of an entity for attending to the whole; the lower level corresponds to the phenomenal. When we switch our attention to the particulars, the subsidiary function of the particulars is suspended and we lose sight of the whole to which we were attending. The logical relation between the subsidiaries and the focal whole gives rise to the semantic aspect in which meaning arises. In Polanyi's thought, "the universe is filled with strata of realities, joined together meaningfully in pairs of higher and lower strata."[2] All of the universe, inanimate, animal, and human, has the structure of a stratified reality.

The relationship between intention and performance follows the same structure. We are focally aware of our intended performance, and our performance is implemented by the integration of subsidiary particulars. When we aim to ride a bicycle we hold that intention focally as we are aware of the particulars of motoric action subsidiarily. The aim to ride a bicycle is the upper level coherence that both relies on and gives meaning to our muscular movements. Knowledge of the physics that our motoric action represents does not enable us to complete the intended action of bicycling. This knowledge is less meaningful when isolated from the intention to perform the action.

In learning to ride a bicycle, even the intention to complete the action does not fully empower us to do so. A gap arises between the intention and performance. The imagination is invoked to close this gap. The imagination thrusts forward to close the gap, invoking unspecifiable clues sensed by intuition. Again the structure follows, the focal action of the imagination relies on and gives meaning to the intuitive sensing and reintegration of clues that work subsidiarily. The spontaneous action of the intuition is powerless without the action of the imagination. The imagination is only successful in completing its aim by relying on intuitive intimations. Polanyi explains that "the imagination must attach itself to clues of feasibility supplied to it by the very intuition that it is stimulating; sallies of the imagination that have no such guidance are idle fancies."[3]

Intuition determines the direction the imagination is to follow, but the imagination is not limited by the integrative workings of the intuition. An area is left open which the imagination summons to achieve an aim that could not be achieved otherwise. The imagination imposes boundary conditions on the intuition. The intuition is subject to dual control: its own integrative properties and the focal intention of the imagination. The imagination's successful action results in the implementation of performance.

The triadic structure of knowing is included in Polanyi's hierarchic schema. We have seen that all knowing relies on the triadic structure in which the knower relies on a subsidiary awareness of some things to attend focally upon something on which they bear. This structure is sequential and made up of a hierarchy of levels. The lower level functions subsidiarily to the next higher level of focal concern.[4]

A sequence of artistic triads sheds light on the triadic hierarchy of levels and the artist's place in society. In creation, the artist works from the triadic structure of knowing between the artist, the world held subsidiarily, and the as yet-unknown artifact. A subsequent triadic level between the artist, the artifact, and the community of interpreters also plays a role. An artist may create in solitude, but the artist attends from his/her world and creates with the intent of community sharing. Creation is not a form of subjective therapy to help the artist cope with life in some way, but is a personal outpouring of form with universal intent. The artist is beguiled, even driven, to create in order to participate with the community of interpreters in the discovery of supra-natural reality.

The actual artistic process is a triadic level that prefaces the social triadic relationship between the artist, artifact, and community of interpreters. On the artistic level what emerges as focal, the artwork, and what functions tacitly are always in a fundamental sense a reflection of or witness to the involvement of the artist with his/her world in an ongoing social process. Creativity is personal—the artist participates in it infused with passion and commitment while rooted in the trust of social companions, tradition, and the acceptance of the community. In the social triadic relationship, the meaning of the artwork relies on the artistic level of creativity but is not determined by this level. While the artist's commitment and encounter with his/her world is an integral particular of the artifact, the meaning of the artwork cannot be discovered by a viewer's attending to this opening level. To do so is to destroy the comprehensive whole of the artifact.

Within the level of the creative process the particulars are integrated by the artist to create the artwork, but the artwork is not explicated by the particulars. A creative discovery is not made by a diligent performance of a previously known and specifiable procedure. The artist has to cross a logical gap by relying on the unspecifiable impulse of heuristic passion, and at the same time be changed by it just as the viewer is changed when engaged with an artwork. An artwork that is characterized by a methodological principle, such as a color theory, is illustration; a painting that cannot be characterized as such lest it lose its meaning is art.

The artist creates with the intent of sharing her/his work, yet s/he cannot force her/his artwork on to a community of interpreters. The artist participates in but does not create the community and is contingent to its acceptance. The artistic level is vulnerable to the authority of the social level and thereby a source of tension in the artist's life. The artist sets out to enrich the world by offering her/his gift of form to humanity for acceptance. To be satisfied, the artist seeks response. The extent to which an artist commits himself/herself to a new vision of truth, separates her/him from others who still look within an old framework. If the community is unwilling to accept the artist's gift, the artist may turn the heuristic passion of originality into persuasive passion.[5] It can be a do or die situation. Vincent van Gogh is a tragic case of an artist whose gift was not accepted. He eventually found no recourse other than death.

Polanyi points out that the two conflicting systems of thought, the artist's and a non-responsive community's, are separated by a logical gap; an artist who relies on one framework cannot demonstrate a proposition to persons who rely on another framework. Meaning is grasped when the viewers approach the artwork with the trust that it means something; they view the artwork sympathetically, open to learning and modifying their old framework. Viewing art can be a converting experience as the beholders accept the meaning set before them. Their former framework will be modified permanently by the artistic creation.

Art's life in society is an imaginative enterprise. The artist's imaginative invoking of his/her world accounts for art's creation; the beholder's continuous imaginative response to the artifact renews its meaning. Art in its making obeys personal self-set standards that the artist accepts. These are rooted in artistic tradition and aesthetic values. Acceptance of art is ultimately grounded in the artist's self-set standards rather than the whim of public acceptance. The labor and perseverance involved in making art tests self-set standards more rigorously than the public does. The community of interpreters, then culture, rely on the self-set standards of the artist in their appreciation of art. This reliance does not determine art's acceptance but protects it from the dictates of public relevance. When a community is moved and changed by art, the framework for acceptance is modified to include consecutive and novel discoveries.

The community of interpreters is a level susceptible to culture. Not everyone is a part of the community of interpreters. This community has no central authority, nor is it a random collection of viewers. The community of interpreters operates from the same premises as the artist, though a multitude of dispersed points exist. The artist is not primarily involved with what scientists or mystics think. The artist is concerned with a new creation that relies on aesthetic value, the opinion of art interpreters, and the belief that they are bound eventually to recognize the truth presented. Even though the new creation may involve a reconsideration of the traditional grounds of art, the artist will still appeal to that tradition as the common ground between himself/herself and the opponents; they in their turn accept this premise. Part of the artistic tradition is its being on guard against suppressing by mistake some great artistic discovery, the claims of which may at first appear meaningless on account of their novelty. The artists and community of interpreters speak with one voice because they are informed by the same tradition and aesthetic values. Any given artistic opinion represents only a temporary and imperfect embodiment of the traditional standards of art.

To accord competence to the decisions of artistic opinion would be meaningless unless a culture accepts art to be, on the whole, true and significant. If culture believes in art, it will accept competent artistic opinion on the whole as valid, even though the final validation of any proposition will always involve some amount of personal responsibility on the part of the individual viewer. Art is grounded in culture, since the artist is grounded in a relationship with her/his culture in creating the artwork; art forms culture because the community responds to the painting and is changed by it.

Reality is a meaningful union of two levels. Pairs of levels join together in a series which forms a hierarchy of levels that relies on the basic logical relation of pairs described above. Each level is made possible and limited by the level below it but is not determined by it. The higher level harnesses the boundary conditions left open by the lower level and is subject to the failure of the lower level. The levels of language are exemplary. The meaning of spoken sounds is found in the spoken words they form, which set boundary conditions to those sounds. The words in turn have their meaning in the sentences they compose, which are regulated by laws of grammar. A style harmonizes the grammatical sentences, which in turn makes a cohesive construction of an idea possible.

Accordingly, the operations of a higher level cannot be accounted for by the laws governing its particulars forming the lower level. We cannot derive a color theory from the colors; a correct use of color theory does not account for style; a style does not determine the content of a painting. The artist relies on artistic rules which govern the conditions of materiality, but these cannot explain artistic meaning.

The structure of levels makes clear to art criticism that analysis of various aspects of art, such as iconography and form, cannot determine the artwork's content. The content is on a higher comprehensive level of the artifact where it relies on and is limited by form and iconography, but is not explicable by them. An analysis of form and iconography can ascertain why an artwork doesn't work and why it falls short of aesthetic value, but only a comprehensive integration of beauty, aesthetic value, and originality can point to an artifact's meaning or content. Form and iconography, constituted by materiality, design, and subject matter, leave open vast boundaries for an artwork's commanding form to harness for an embodiment of meaning. The possibilities for the creation of something new are enormous. The commanding form relies on formal properties in its search for form, and it integrates these principles so that they work together and embody content. The logical relation between the lower level of form and iconography and the higher level of content gives rise to the artifact's joint meaning.

Form and iconography determine the direction the commanding form will follow, but the commanding form is not fully limited by their materiality and conceptions. The commanding form demands novel integrations of these specifiable elements in order to achieve what has not existed before. The commanding form's successful embodiment of content results from imaginative integrations of these elements. The principle of "dual control" allows for novel re-integrations of natural reality so that supra-natural reality may be made tangible.

First, then, our tacit integration of particulars on the lower levels changes the particulars, but it does not upset the laws by which they are governed on their own level. When we recognize a clock for what it is and does, the apparently random parts do not change but take on meaning as a timepiece. Likewise, what could be perceived as random marks on a painting take on the form of a landscape when integrated; the parts do not change though their appearance does.

Second, the higher level is the authoritative level, since it is only in our perception of the higher level that we can see what the items of a lower level fully mean. For instance, the parts of the clock are recognized as a particular mechanism that keeps time. Or, the random marks of the painting are transformed into a meaningful whole that we claim to be a landscape. Usually the comprehension of the higher level is irreversible. Once a clock is known for its purpose, viewing the parts of the clock as less meaningful entities is virtually impossible. When we analyze the parts, we hold tacitly the knowledge that they are parts of the clock. So too, once we recognize a form in a painting, viewing it another way is difficult.

All of the consecutive levels in the universe are real. Indeed, following Polanyi's definition of reality, the higher levels are more real than the lower levels. Since the higher levels are more comprehensive, they have more potential undiscovered implications. The higher the level of an entity the greater are the imaginative powers required to comprehend its successively more intense and complex integrations. Physics, which deals with patterns of inanimate nature, is less complicated than biology, which is concerned with living beings. The subjects of physics have fewer consecutive levels than biology. The more levels involved in the coherences, the more profound will be its meaning and our understanding. This is due to the higher level being an enlargement that gives meaning to the lower level. The structure of knowing is the same for all levels, but the more comprehensive a matter, the more is demanded of our powers to understand.

According to our definition of supra-natural reality as the natural world revealing a deepened meaning, supra-natural reality belongs to a higher level. Supra-natural reality is grasped through clues innate to the lower natural level. When supra-natural reality is apprehended, it breaks us out of former ways of understanding every level of reality. In accordance with the principle of "dual control," the particulars on every level are true to the laws on their level and are elements of a higher comprehensive entity as well.

The higher level emerges in the context left open by the lower level. Supra-natural reality summons the lower level by responding to boundary conditions which enable the lower level laws to work for a higher principle. Boundary conditions are not imposed; they open a lower level to novel possibilities of a deeper meaning. Meaning is achieved through our seeking on a lower level clues that point to a greater comprehensive whole.

Art embodies clues from natural reality that when imaginatively integrated present a higher level of supra-natural reality. Art, then, can be said to be a greater comprehensive entity than can be grasped on the natural level. It is a more complex manifestation of meaning than indication. As a self-giving integration, it carries the viewer into itself. The viewer surrenders the self in an imaginative act to grasp supra-natural reality.

The contemplative pours the self intuitively into natural coherences and discovers supra-natural reality. The contemplative resides in the present; the viewer of art thrusts toward future manifestations of meaning. The contemplative and artist grasp deeper coherences of reality than are ordinarily experienced. The contemplative immerses the self into a reality that does exist in nature to comprehend a deeper now than language can articulate fully. The need for symbol and metaphor arises for its expression. The artist surrenders the self into natural reality and re-integrates it imaginatively through form. The viewer is carried imaginatively into the reality of the artifact to grasp art's presentation of a deeper future. The artifact's reality relies on nature but does not in itself find its full meaning in nature. Natural reality can only hold the potentiality for supra-natural reality to be recognized by the contemplative or incarnated in works of art.

Religion, too, is a tacit integration of clues to a yet deeper level of meaning. The divine level summons and responds to the human level. Human beings experience a never-ending call to discover progressively deeper coherences of meaning. The transcendent reveals itself immanently. The human level makes possible and limits the divine level but does not determine it. Because human beings are created in the divine image, we are made to know the divine; because we are limited by finitude, our perspective on the divine is restricted. The human level also breaks free from the divine level's call. Humanity is empowered with free will and succumbs to sin. To reach the divine level requires leaps of the imagination and intuitive sensing that opens new vistas of understanding of our world. Through committed seeking, stimulated by beauty, we can, if only partially, discover the Supra-Natural meaning that transforms our understanding of natural reality.

Comprehensive realities known in art and religion beckon to us—they are beautiful and they carry us away. They embody many of the deeply rooted tacit elements that constitute a person as a social being. In the act of aesthetic and religious knowing, the viewer/worshiper dances with beauty in order to understand successively more profound coherences of supra-natural reality.

NOTES

1. Michael Polanyi, *TD*, p. 34.
2. Ibid., p. 35.
3. Michael Polanyi, CI, p. 104.
4. Michael Polanyi, *KB*, p. 186.
5. Michael Polanyi, *PK*, p. 159.

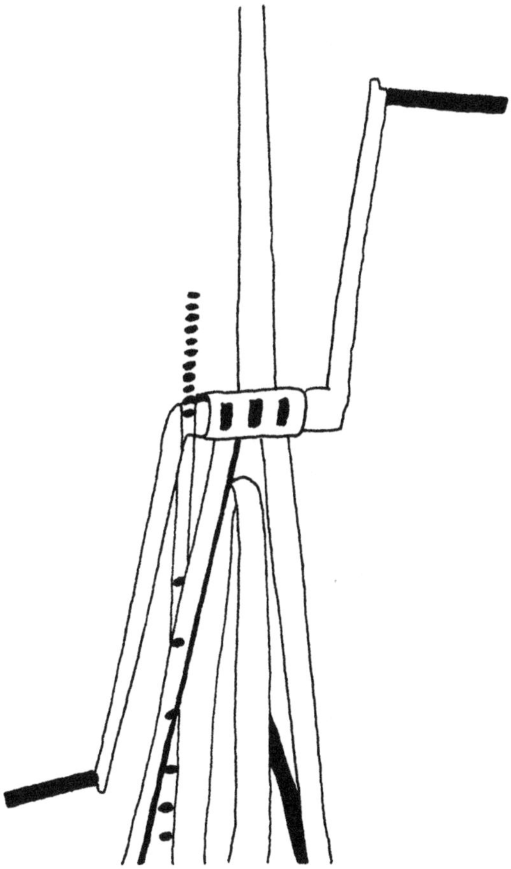

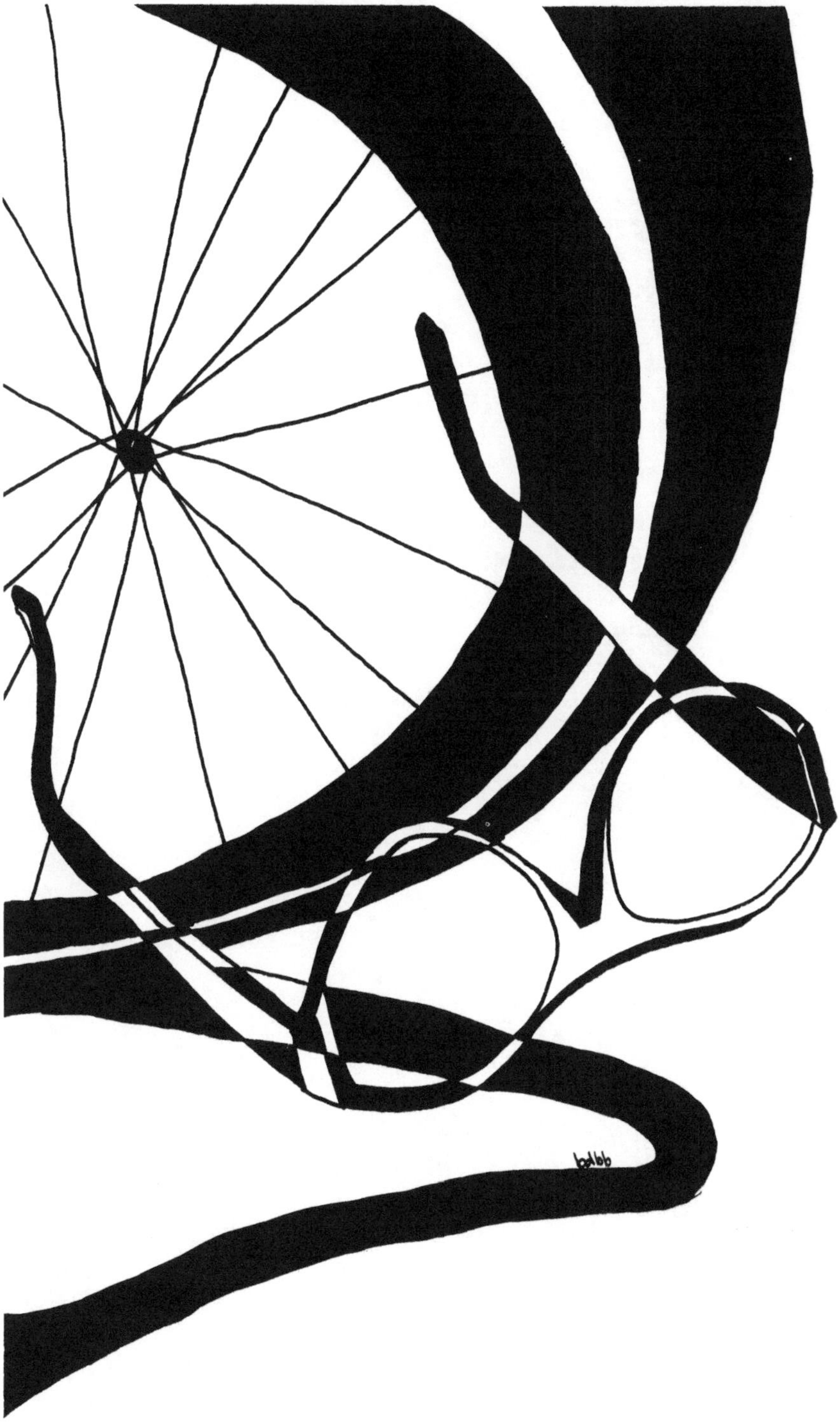

Chapter Six

The Quest for Meaning: A Theological Dimension

Art and religion, like symbol, metaphor, and contemplation, are self-giving integrations. They enlarge our basic structure of being in the world by "drawing us in rather than cutting off our historical experiences as persons."[1] The ordinary from/to structure of knowing that projects away from the self to the focal object is enlarged to include the self by a pouring of the self into the comprehensive meaning. Personal participation and indwelling of clues is involved in all cases of knowing, but only self-giving integrations carry the self into the focal object, thereby giving our socio-historical being embodied expression.

The visual arts rely on beauty, aesthetic value, and originality for their self-giving expression of content. Art carries the self into its visual embodiment. Religious merit, which includes beauty, religious value, and convention, call the religious believer to embody religion into rites of liturgy or worship. Indication, symbol, metaphor, contemplation, and art are all elements combined in varying ways to give liturgy its form.

Matters of religious value include (1) faith, which calls us to know even in the absence of certainty; (2) assertions of belief, which include doctrines or religious teaching, creeds, factual knowledge, and personal experiences of the divine; and (3) performance, which speaks to how we live in light of religious beliefs through morality, service, and religious expression or liturgy. Style suffuses religious frameworks such as Buddhism, Christianity, Judaism, and religion's form of worship, such as charismatic, sacramental, evangelical.

Religion relies on convention rather than originality because religion presents a comprehensive framework which endures through generations. While our understanding of religion's framework changes, we do not in turn re-write religion's basic assertions.

In liturgy's attempt to embody aspects of religious merit through its enduring form and iconography, it provides an opportunity for the worshiper to experience matters of religious beauty and value. Liturgy extends the imagination beyond ordinary experience and provokes the imagination of the worshiper which evokes a deeper coherence to life. Liturgy induces and expresses our passionate search for God.[2] And it acknowledges and appreciates the God we have come to know.

Polanyi approaches his interpretation of liturgy along the same lines as his explanation of art. The fundamental difference resides in liturgy's essential lack of originality. "The theme of commemorations is conventional and its forms are traditional; but [like art] the themes are important and the forms are often deeply moving."[3] Through recollecting events that touch the depths of all human experience, religion, using recurrent traditional forms of liturgy, is able to maintain a comprehensive framework that lasts through the centuries. When art in its novelty presents profound and intrinsically significant events, it is an artistic classic. Its meaning, too, lasts through the ages.

Both art and religion offer to our imaginations a problem that is ultimately insoluble. They break through our accepted frameworks of thought and require a high degree of imaginative effort to move us beyond our present locations. For this reason they often meet with resistance on our part. We are called to surrender ourselves into an heuristic vision in which interpretation is difficult. We find staying in our present location much easier than picking up and moving into a journey that leads beyond familiar horizons. The tendency is to cautiously hang on to elements of familiarity: a creed, a ritual, a style, a natural form. Rather than trust the imagination's lead into the comprehensive meaning of the liturgy or artifact, we focus on subsidiary elements that we know. Therefore, the meaning is dismembered, we are not carried away, and the experience is rendered less meaningful.

1. Detachment and Self-Giving Integrations

In Polanyi's thought, the key to what makes art and liturgy distinct from other forms of knowing is not their self-giving nature but their detachment, because art and worship are profoundly unique experiences. They break into the course of our everyday lives and set our imaginations free from ordinary experience. They stand outside of everyday space and time; art speaks of a single moment within its own space, liturgy speaks of another time and space. Our submission to liturgy consists, as in the case of art, "in being overcome by it and carried away into its own sphere, away from the sphere in which we lived a moment ago and to which we shall presently return."[4]

According to Polanyi, we enter into and grasp the meaning of an otherwise detached artifact or liturgy through self-surrender. The structure of self-giving integrations are distinctive because they enable us to participate in a detached event. Detached events are cut off from ordinary experience, and we do not mistake them for everyday occurrences. By surrendering into them we are able to grasp the "wholly other" events of art and liturgy.

That art and liturgy are detached appears to be Polanyi's primary point. He is concerned with the structure of self-giving integrations insofar as they explain our ability to enter into a "wholly other" event. In other words, self-surrender is a subsidiary which makes possible our recognition of meaning in an otherwise detached event. But self-giving integrations are not a decisive quality in Polanyi's discussion of semantic relations.

Polanyi describes detachment as arising through artificial coherences that "frame" art and liturgy metaphorically. This framing harmonizes incompatible components into a comprehensive entity, while it detaches the person from the artifact/liturgy. Polanyi's theory relies on our ability to integrate disparate elements into a whole which has a quality of depth absent in each distinct part.[5] Through this process novel traditions are brought forth which require feats of the imagination to grasp.

The embodied harmony does not resolve the incompatibles but holds them in a heightened tension that presents an emotionally charged entity. We, like the creator, cannot solve the tensions, we can only wrestle with them. Polanyi cites a few examples of aesthetic incompatibles: the flat canvas is incongruous with perspectival depth; an artificial pattern of form clashes with iconography; liturgy moves a person into a surrounding s/he does not daily inhabit. The fusing of incompatible elements detaches art and liturgy "in the sense that they do not predict any observable facts."[6] This is because as facts of the imagination they are out of context from ordinarily experienced time and space, contends Polanyi.

Considering this view of Polanyi's in light of the hierarchy of levels, let us return to our consideration of the clock. The clock is made possible and limited by physical parts though not determined by those parts. The operational principle of the clock governs the functioning of the parts, but it cannot predict the failure or success of a part to function. The operational principle is concerned with the overall functioning of the clock, not each individual part. Yet the clock is not detached from the parts. It summons and transforms the parts into something that they could not otherwise be.

Art and liturgy affirm natural reality while employing its parts to mean something larger than they mean in themselves.[7] Art and liturgy are gradients of reality that require an increasing measure of imagination to grasp. Following the model of the clock, Polanyi is right; art and liturgy cannot predict facts of observable reality. But they are not wholly detached from natural reality; they transcreate[8] reality into a presentation of supra-natural reality. Art and liturgy are not out of context; they present a deeper context of time and space. They harness ordinary experience and extend a more unified reality than can be otherwise known. Sabbath worship is not wholly detached from our workaday lives but transcreates the everyday into a meaningful whole. The art quilt, for example, has been a powerful form of art in recent years because it is intensely *attached* to our daily lives. People are especially interested in the art quilt because the quilt is familiar as a blanket and a symbol of our grandmother, yet it is unfamiliar in its formal presentation.

Art and liturgy, then, do not present a wholly detached sphere for us to enter into while we leave our present sphere. We pour ourselves into the artifact or liturgy by carrying our present location into the experience. In so doing, a comprehensive event takes place in which we jointly participate with the artist, the artist's community, and our community, or in liturgy, with God and the religious community. A deeper sphere of our ordinary sphere is experienced so that when we return to the everyday we do so with an expanded and profound understanding of the ordinary. What makes art and liturgy distinct is that they are simultaneously imaginatively detached from, yet attached to, ordinary experience. They create form which embodies a transcreation of reality into supra-natural reality.

Our ability to experience this supra-natural reality pivots on the metaphorical quality of art and liturgy that is self-giving. The life of our imagination is, in part, the tenor that bears on the vehicle of art or worship. Art and liturgy bear back on our imaginations, thereby embodying our imaginations in the artifact and liturgy. Art and liturgy are indeed facts of the imagination. Finally, the self as a socio-historical mindbodily being in the world is poured into these facts of the imagination. Weismann's telling of creation demonstrated that our tacit location takes on an intense focal awareness. The self is carried away and possessed into an extraordinary experience of art and liturgy.

The imagination is not detached from reality. If it were, its achievements would be illusion or fallacy. The imagination is our faculty to seek and discover our world and its meaning. It quests for beauty. It is imbued with intuition, passion, and all that we are mindbodily. Art and liturgy carry us into an extraordinary structure of knowing so that our experience as mindbodily beings in the world may be enlarged and deepened. I believe that the "into" structure of self-giving integrations extends Buber's "I-it" relations between the knower and the known into "I-thou" relations.

Detachment and attachment are held in an insoluble tension of harmony. Detachment, I contend, is an incompatible particular of self-giving integrations that harmonizes with attachment. Through the self-giving structure of art and liturgy, their extraordinary presentation of space-time is fused with our and the creator's ordinary experience of space-time.

Yet the distinctiveness of art and liturgy are due to more than the two features of metaphor: self-giving integrations and the fusing of incompatibles. Metaphor is an aspect of art and liturgy that mixes with the other features of iconography, form, beauty, and originality or convention. This intermixing requires a high degree of personal participation in order to be sustained. The viewer/worshiper is called to imaginatively enter into art and liturgy's transcreation of reality into supra-natural reality.

2. Attachment and Detachment: Incompatibles of Time-Space

The aspect of metaphor that fuses disparate elements into a joint meaning produces a frame which detaches art and liturgy from everyday experience. The other aspect of metaphor, its self-giving nature, profoundly attaches art and liturgy to the everyday. These two elements of detachment and attachment lend the metaphoric quality of art and liturgy their power when they work together. Attachment and detachment are most profoundly evident in our experience of space and time. Because space and time are central to our mindbodily experience in the world, so too are their workings within art and liturgy.

Polanyi notes that art captures one moment of time. It is detached from experience because it stands outside of time. This is similar to William Poteat's view that visual art is an image that seizes, arrests, and holds the eye. The particulars appear to be simultaneously co-present before the eye. Art as an object is changeless and endures time. To change even a single element of an artifact destroys it or makes it into something new. This is true even in cases of contemporary participatory art. Since participatory art demands to be changed by the viewer or its "happening," the changing aspect of the artifact is a particular of the work of art. When the change does not occur, the artwork loses its meaning, and in this way the artwork is changeless. The artifact must be true to its commanding form which does not change even when its particular includes change.

The essential changelessness of art detaches art from ordinary time. But art, like all reality, can only be experienced temporally. Polanyi omits this feature in his aesthetics. The temporal tension that art heightens is an enduring configuration fused with a dynamic and progressive event. The detached artifact that endures time is integrated with the temporal self, and its meaning can only be grasped over time.

The case is similar in liturgy and religion. Religion's enduring comprehensive framework and liturgy's recurrent traditional forms are akin to the changelessness of the artifact. Our affirmation of a religious framework and our participation in liturgy occur temporally through our socio-historical mindbodies. A crucial tension is established that enlivens and enlarges religion and worship. When we omit one of these temporal incompatibles, the essence of transcreating possibilities are lost. When we focus on the structures of convention, religion and liturgy become overly routinized and meaningless. When we focus on new particulars that arise through our existential context of knowing and being, religion and liturgy become relevant without depth. Religion and liturgy become subject to the "changing of the wind." The meaning dies as quickly as it rises.

The spacial component in visual art is much greater than a flat canvas that conflicts with perspectival depth. This incompatibility, of course, falls apart with abstraction (and it doesn't apply to sculpture and architecture). Polanyi handles the problem by stating that the incompatibles of abstract art are dismembered elements of natural form that "make no assertions...their integration places them directly into a field of the imagination wholly detached from the personal concerns both of author and public."[9] I find this argument confusing and inconsistent with Polanyi's overall scheme of personal knowing. Polanyi confounds non-representational art's requirement of immediate indwelling with detachment. Space, like time, becomes detached from ordinary experience. As an object, art is as detached from our experience like any other object outside of ourselves: a rock, a flower, a person. Polanyi successfully asserts that no reality can be known in a strictly detached manner of observation. All of reality is known personally. Art is no exception.

The space of the plastic arts is distinct because it makes space visible. The plastic arts organize space in order to make it into something it could not be otherwise, visual. Natural space makes visual art possible, but visual art limits natural space to the visual realm in its concrete expression. Visual art is transpicuous,[10] since it incarnates space into an imaginatively charged and meaningful visual entity. It is not wholly detached but successive and distinctly meaningful in its visibility. The infinite and dynamic space of ordinary experience and the concrete visual space of the plastic arts are fused into an unresolvable tension. The result is a complex spatio-temporal configuration that is tangible to the eye and self-giving.

Religion integrates infinite, all-encompassing space with our limited existential space. Infinity lends vital meaning to our incarnated world. Liturgy integrates private space with communal space. Within Christian liturgy, individual bodies are affirmed, while at the same time they function as the communal body of Christ. The Christian experiences liturgy with a physical discomfort that is rooted in this tension. When communal space is disregarded, worship becomes isolated and privatized. When individual bodies are negated, existential rootedness is lost and liturgy loses its depth. In either case, the worshiper is not carried away, and liturgy becomes an empty shell.

Detachment and attachment are tensely fused within the self-giving integrations of art and liturgy. When the self is surrendered into the artifact, our mindbodily socio-historical being is fused into the otherness of the artifact. Metaphoric integrations fuse two sets of incompatible self-giving particulars. The first set is intrinsic to the artwork. Its style suffuses form and iconography with all that the artist brings mindbodily into the artifact. The second set includes all that the viewer pours mindbodily into the work of art. A dynamic tension between, on the one hand, the artist and the artist's world and, on the other hand, the viewer and the viewer's world, is experienced. The result is an harmonious, mysterious, and passionate event highly charged with meaning.

That Polanyi mistakenly discerned that art and liturgy's key distinction are in their being detached, I believe, results from his own detachment from them. He did not have enough direct experience with art or liturgy to understand either. Skill and connoisseurship are as integral to appreciating art and religious understanding as they are to scientific discovery and medical diagnoses. The knower must dwell imaginatively in the clues of art and religion in order to break into their respective presentations of meaning. Tradition, trust, and intuitive sensing enable us to surrender ourselves imaginatively into the framework of religion, into worshipful participation, and into the artifact. This moves us beyond from/to knowing into an enlarged and enriched meaning that transcreates ordinary experience.

3. Art: Religious and Theological

Ordinarily we comprehend reality partially. Religion attempts to unify reality and invites us to comprehend reality more wholly. Art helps us along the way. It integrates disparate parts, and it makes visible content that cannot be expressed in any other way. In so doing, visual art transcreates reality into a visible presentation of supra-natural reality. Art's supra-natural reality reveals a profound coherence of meaning not ordinarily grasped. The intention of art, to reveal ever deeper coherences of meaning which lie beyond the realm of articulation, emulates the intention of religion which seeks coherences of the whole of our world. Religion provides an interpretive framework which connects our life together. In religion's view, all of life hangs together and we are called to experience life whole. Ordinary experience is fused with extraordinary coherences. Attachment and detachment are integrated into an harmonious whole, though their tension is unresolvable. Art, too, holds together elements of attachment and detachment but in a less comprehensive way. Yet, in this attempt to fuse disparate elements, including ourselves, into an insoluble heuristic vision, art appears at heart religious.

Polanyi, I suggest, sensed the religious nature of art. The framework of personal knowing, the theory of levels, and the metaphoric quality of self-giving integrations and harmonized incompatibles are evidence of this. He just did not allow his imagination to ferret out the path that his intuition sensed.

Following Polanyi's lead, I discussed religion, liturgy, and art jointly. They are closely allied, but they are not the same thing. Remembering that the higher level enlarges the lower level by summoning and transcreating it, we may recognize religion as a higher comprehensive entity than liturgy. Liturgy is a higher, more complex fusing of consecutive levels than art. Religion harnesses liturgy for its expression; liturgy is the heuristic enquiry of religion. It relies on art, metaphor, symbol, contemplation, and indication in its pursuit to express and stimulate the desire to know God. Liturgy is a more complex integration than art.

Art is, however, indispensable to the clarification of religion, of liturgy, of culture because it can make the intangible tangible. Religion and liturgy need art in their expression, but art potentially has its independent existence outside of religion and liturgy. It does not necessarily rely on them. Art presents a complex self-giving integration that guides us in our pursuit to understand religion, culture, and all that we are mindbodily in the world. It incarnates the meaning that motivates and underpins our being. Is this theology?

Polanyi defines theology as a

> context of worship and [it] must be understood in its bearing on it.... Theology as a whole is an intricate study of momentous problems. It is a theory of religious knowledge and a corresponding ontology of the things thus known. As such, theology reveals, or tries to reveal, the implications of religious worship, and it can be true or false, but only as regards its adequacy in formulating and purifying a pre-existing religious faith.[11]

Theology helps the community of believers to understand who they are, what they believe, and what they practice in liturgy. Insofar as art helps the community of believers to understand themselves in relation to their faith and helps them surrender into worship, it can be said to be theological. On the whole, however, art is not concerned with formal aspects of theology but is an expression of what senses, implements, and seeks that theology. The theological dimension of art invites us to seek God when art implements our intention to find an expression of God within art's event.

Theology frames and stabilizes religious beliefs that are held by those who dwell in that religious framework. To understand a given theology we need to dwell in the religious framework. Much religious content cannot be articulated by theology. Art can present the theological meanings that elude verbalization. Yet, theological art is created from the artist's dwelling in a religious framework and the theology art presents is grasped by those who dwell in aesthetic and theological frameworks.

The meaning of the artifact cannot be fully discovered through specific interpretive frameworks. These frameworks are subsidiary tools indwelled by both the artist and the viewer. In art appreciation, religious frameworks and theology are held tacitly. As a part of the self, they function as subsidiary elements that are embodied into the artifact by the artist or the community of interpreters. Ultimately, art breaks us through our frameworks to reveal a deeper meaning than we grasped before. Art transforms our frameworks when we approach it with an heuristic passion and submit to its demand that we change our view of reality into a new and deeper vision.

Since art cannot be fully described, it does not provide a wholly stable belief system. It is an unstable coherence that does not rely on observable fact or a factual basis of experience. Instead, it is an imaginative performance visible only to the imagination. Art speaks its own language. We cannot make it speak the language of our location, since it expresses a particular idea to be enjoyed in itself. What we may do is create a conversation between the artifact's language and our own. In this process meaning is grasped. In art's claim to be received for its own sake, it presents its intrinsic intelligibility that may or may not be theological.

Alexander Thompson's description of the Apostolic witness is instructive in understanding how art may be theological. He points out that Christian theology draws upon the biblical witness and its exegesis. In the Christian faith, "we dwell in the biblical witness and attend from it to the vision of God which it was the Apostles' intention to convey."[12] The Christian dwells not merely in the explicit statements of the Apostles but also in the tacit, informal element which is passed on as a "living" tradition through the generations. Since we are dealing with a knowledge of God, a reality that far outruns our understanding, a reality that manifests itself in surprising and unsuspected ways in the future, we can state that the Apostolic witness points to more than it can tell. "This allows for a deepening and developing tradition in theology where new aspects of the once and for all revelation in Christ are brought to light but always in coherence with what was disclosed in the Apostolic witness."[13]

Following Thompson's lead, I suggest that theological art presents tacit and informal elements of religious witness that cannot otherwise be expressed. Theology makes assertions about an unnameable and imageless reality and thus is limited in its expression. Religious meaning surpasses our articulate powers. Art can express a deeper content than discourse, though it, too, is limited by its materiality. God cannot be imaged, only pointed to and experienced. Art can be an eventful clue to God.

There are many occurrences in the Bible where the writer moves into poetry to express a reality that transcends articulation. The annunciation story in the Gospel of Luke is exemplary. Luke begins with a discursive account of what happened:

> In the sixth month the angel Gabriel was sent by God to a town in Galilee called Nazareth, to a virgin engaged to a man whose name was Joseph, of the house of David. The virgin's name was Mary....(Luke 1:26-28ff, NRSV)

The account indicates the basics of the story. Following in the same chapter Luke breaks into poesis to express Mary's response:

> 'My soul magnifies the Lord,

and my spirit rejoices in God my savior,
for he has looked with favor on
the lowliness of his servant.
Surely, from now on all generations
will call me blessed;
for the Mighty One has done great things for me
and holy is his name'....(Luke 1:47-49ff, NRSV)

Here we are invited to enter into Mary's world. We do not just learn about Mary but participate in her relationship with God.

Theological art presents a reality that surpasses our understanding, while it is itself a reality we cannot totally comprehend. It is a powerful yet unstable expression that points to more than it can tell, while at the same time it tells more than we can know. Appreciation of the theological dimension of art opens the doors to greater religious meanings than we indwell or anticipate. It heuristically breaks us out of old patterns of meaning and being. While theology explains and can break open religious frameworks, art breaks us into those frameworks. Our being embodied in the artifact calls us to surrender into it and to break into a new being. When theology is involved, that new being is impregnated with religious meaning.

Art can present theology through an artist's dwelling in a religious framework. In this case, that art can be classified as an intentional theological art. Matters of religious merit function subsidiarily to the commanding form. Or an artist that does not dwell in a religious framework may dwell in a community of faith. The artist's trust in and reliance on that community enables the artist to integrate theological meaning intentionally (not focally) into a work of art. Or art can be deemed theological within the context of interpretation. In this case, the artist's intent or indwelling may be irrelevant. Elements of religious value are subsidiarily integrated into the work of art by the viewer but perhaps not by the artist. In the self-giving integration, aspects of religious merit are embodied into the artifact's content. Not that the viewer fabricates a meaning the artwork does not possess, but the religious framework integrated into the artifact deepens the aesthetic experience.

The task that understands art theologically accepts the religious community's self-set standards of religious merit, including beauty, religious value (faith, assertions of belief, performance), and convention. These standards are integrated with the self-set standards of aesthetic merit which include beauty, aesthetic value (style, iconography, form, intrinsic interest), and originality. Criteria of religious and aesthetic merit function jointly in a communal discernment of the theological dimension of a work of art.

Theological content is not determined by the criteria that religious and aesthetic merit gather. Criteria of religious and aesthetic merit are assessed by the artist, theologian, and the community of interpreters so that they may discover through them a more meaningful and valid discernment of theological art. Intellectual passions imbued with religious and aesthetic values allow the viewer to distinguish those criteria that are of interest to their concern. The viewer surrenders into the artwork from a religious and aesthetic framework in faith that the work of art bears on reality. Theological meaning arises from the subsequent witness of viewers who experience and then share a profound religious understanding of our everyday experience.

The community of interpreters jointly apply the criteria which contour religious and aesthetic merit in order to discern the theological dimension of art. This joint application protects the community from valuing an artwork that is irrelevant to theology or lacks contact with reality. To the extent that the viewing community dwells in religious and aesthetic frameworks, these criteria of merit can be appreciated as having a bearing on the theological dimension of art.

In the end, it is the community of faith who is called upon to affirm an artwork as theological. The theological dimension relies on the community's passionate response to the work of art that can never be dispassionately defined. The community's acceptance of a theological dimension protects art from being misinterpreted or forced into a fallacious meaning.

NOTES

1. Carl Phillips Mullins, "Polanyi's Analysis of Art: A Critical Appraisal," unpublished paper delivered at the Centennial Conference on the Thought of Michael Polanyi, Graduate Theological Union, Berkeley, Cal., 15-16 March 1991, p. 10.

2. Michael Polanyi, *PK*, p. 282.

3. Michael Polanyi, "Visionary Art," in M:P, p. 14.

4. Ibid., p. 21.

5. Polanyi, "From Perception to Metaphor," in M:P, p. 19.

6. Polanyi, "Acceptance of Religion," in M:P, p. 1.

7. Polanyi, "From Perception to Metaphor," in M:P, p. 23.

8. The term "transcreate" may be found in *Webster's New International Dictionary*, 2nd ed. (Springfield, Mass.: G. C. Merriam & Co., 1960) and means: to transmit through creating. Art transmits supra-natural reality through creating novel coherences of natural reality.

9. Polanyi, "Visionary Art," in M:P, pp. 10-11.

10. The term "transpicuous" may be found in *Webster's New International Dictionary* and means: to see or look through.

11. Polanyi, *PK*, p. 281.

12. Alexander Thompson, *Tradition and Authority in Science and Theology: With Reference to the Thought of Michael Polanyi* (Edinburgh: Handsel Press, 1980), p. 108.

13. Ibid., p. 108.

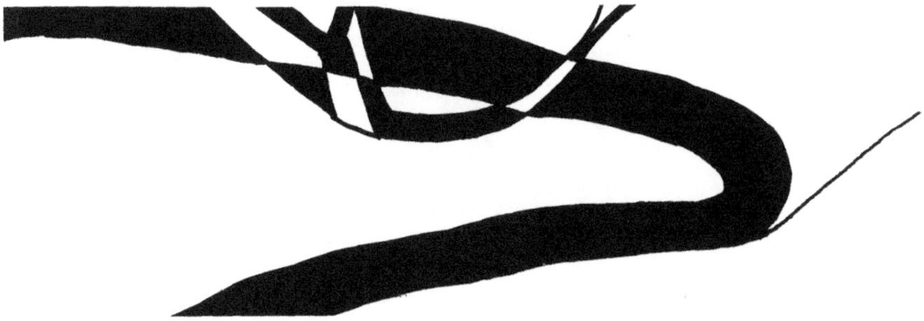

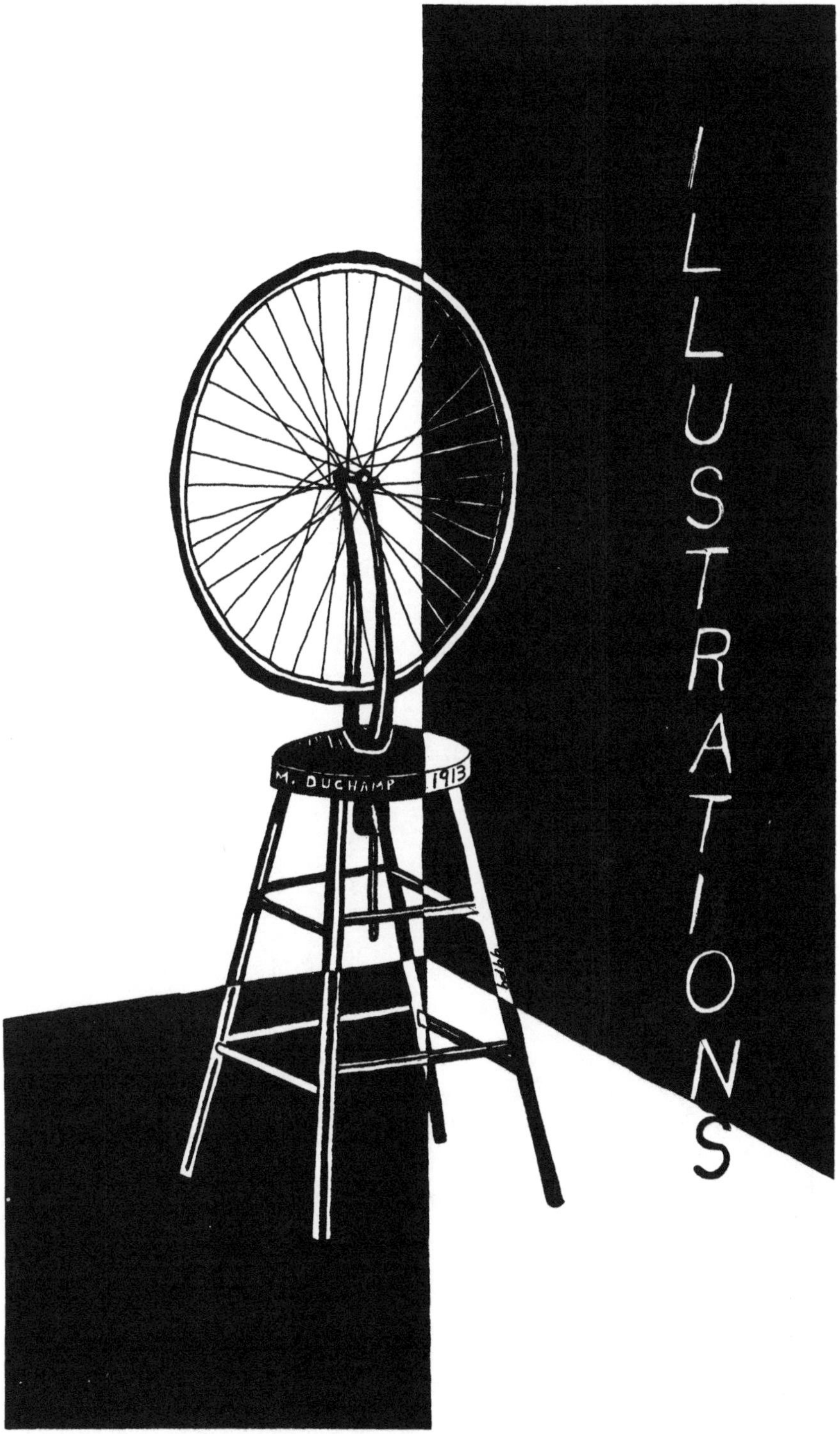
ILLUSTRATIONS
M. DUCHAMP 1913

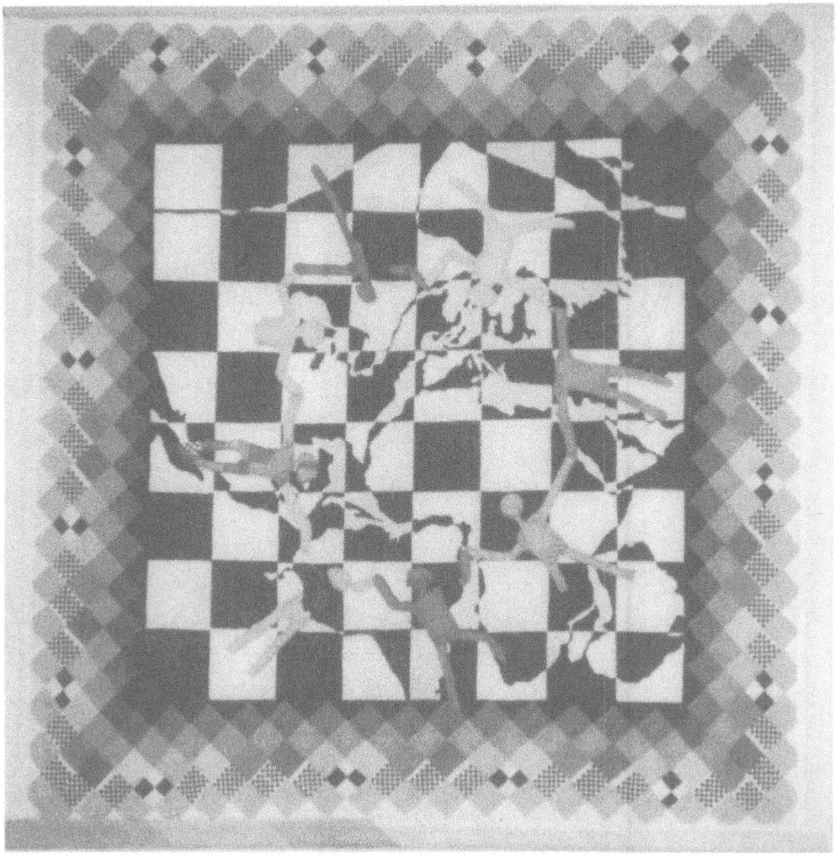

1. Dee Baumgarten, *World*, 1986, quilt. Private Collection.

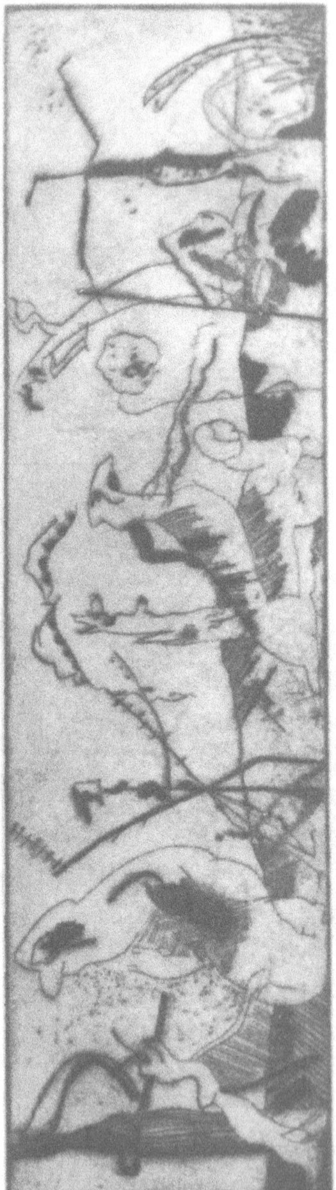

2. Dee Baumgarten. *Dance of the Twigs*, 1976, drypoint.

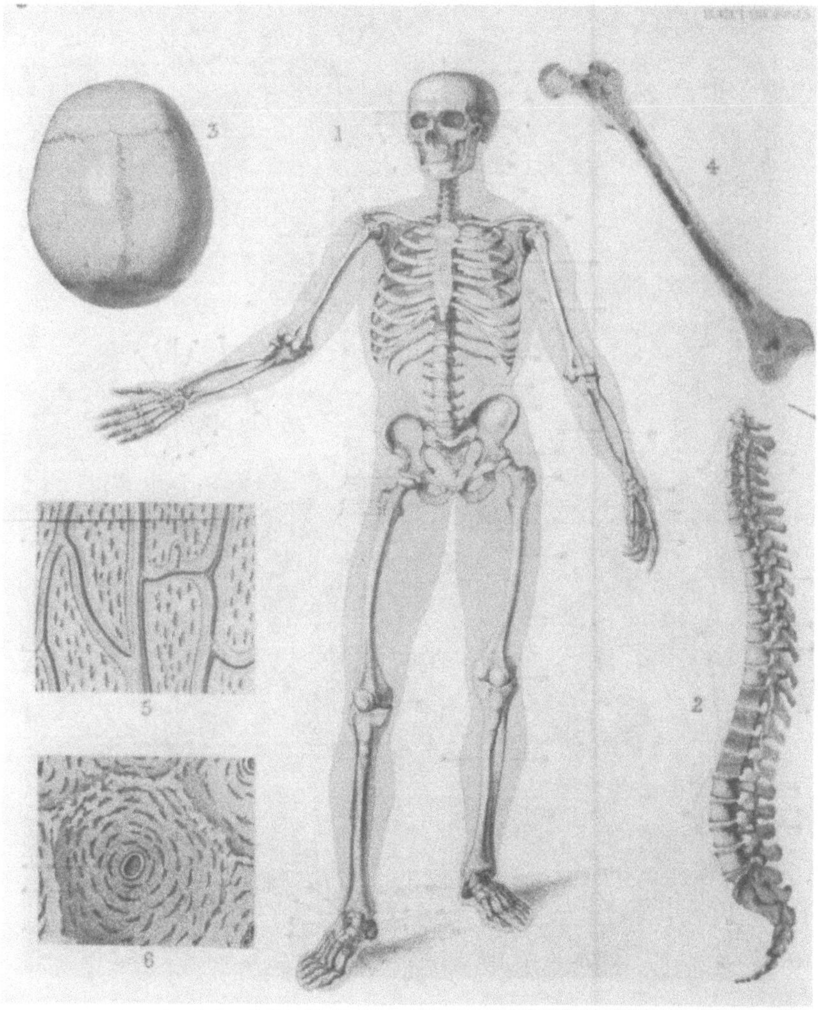

3. W.A.K. Johnston, *The Bones*, 1857, engraving. William Turner and John Goodsir, *Atlas of Human Anatomy and Physiology* (Plate 1).

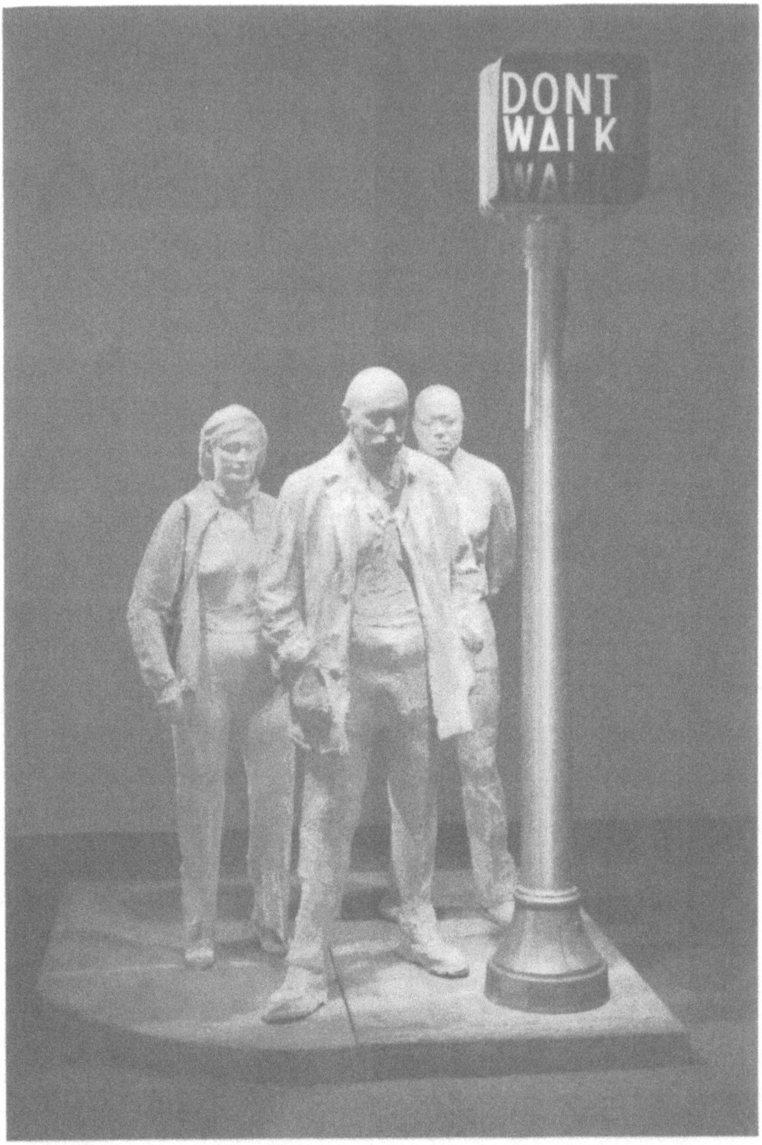

4. George Segal. *Walk, Don't Walk.* 1976. Plaster, cement, metal, painted wood and electric light. 8 feet x 6 feet x 6 feet. Collection of Whitney Museum of American Art. Purchase, with funds from the Louis and Bessie Adler Foundation, Inc., Seymour Klein, President, the Gilman Foundation, Inc., the Howard and Jean Lipman Foundation, Inc. and the National Endowment for the Arts.

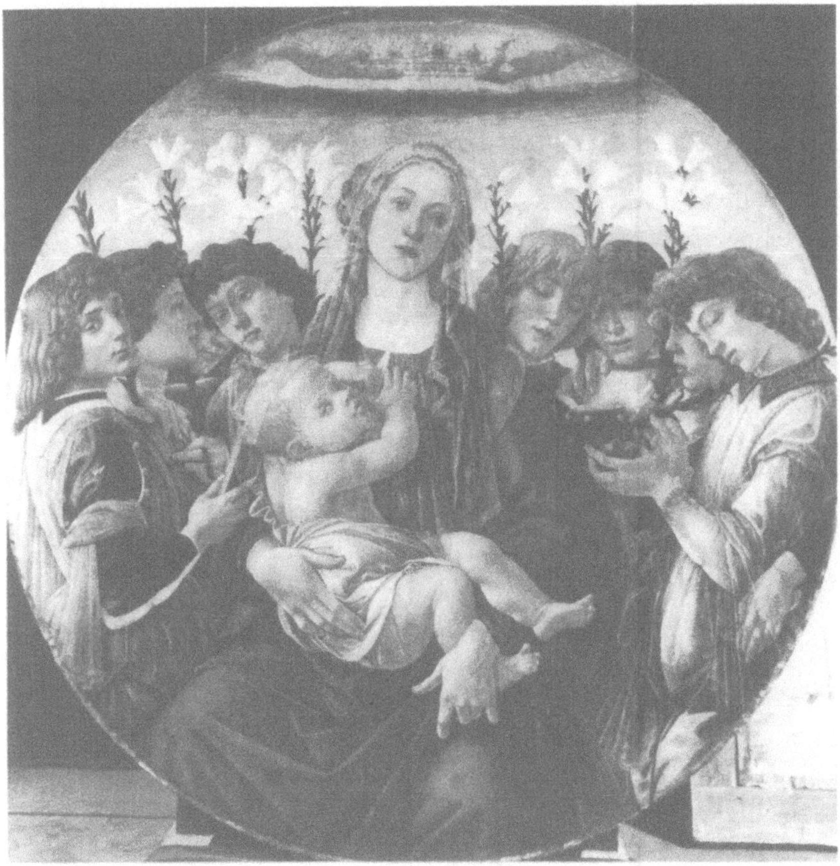

5. Sandro Botticelli. *Madonna and Child with Singing Angels*. ca. 1477. Gemaeldegalerie, Staatliche Museen, Berlin, Germany. Foto Marburg /Art Resource, NY.

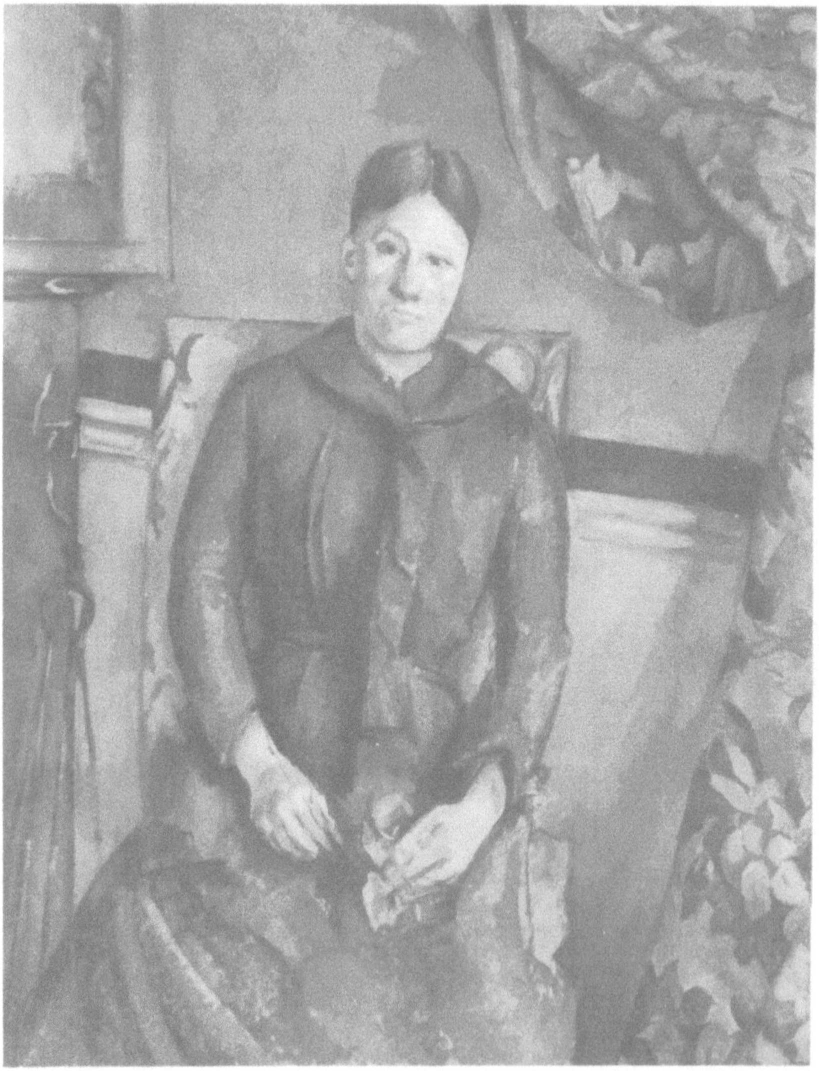

6. Paul Cézanne, (French, 1839-1906). *Madame Cézanne in a Red Dress*. (Hortense Fiquet, born 1850), oil on canvas. Metropolitan Museum of Art. Purchase, Mr. and Mrs. Henry Ittleson, Jr. Fund, 1962.

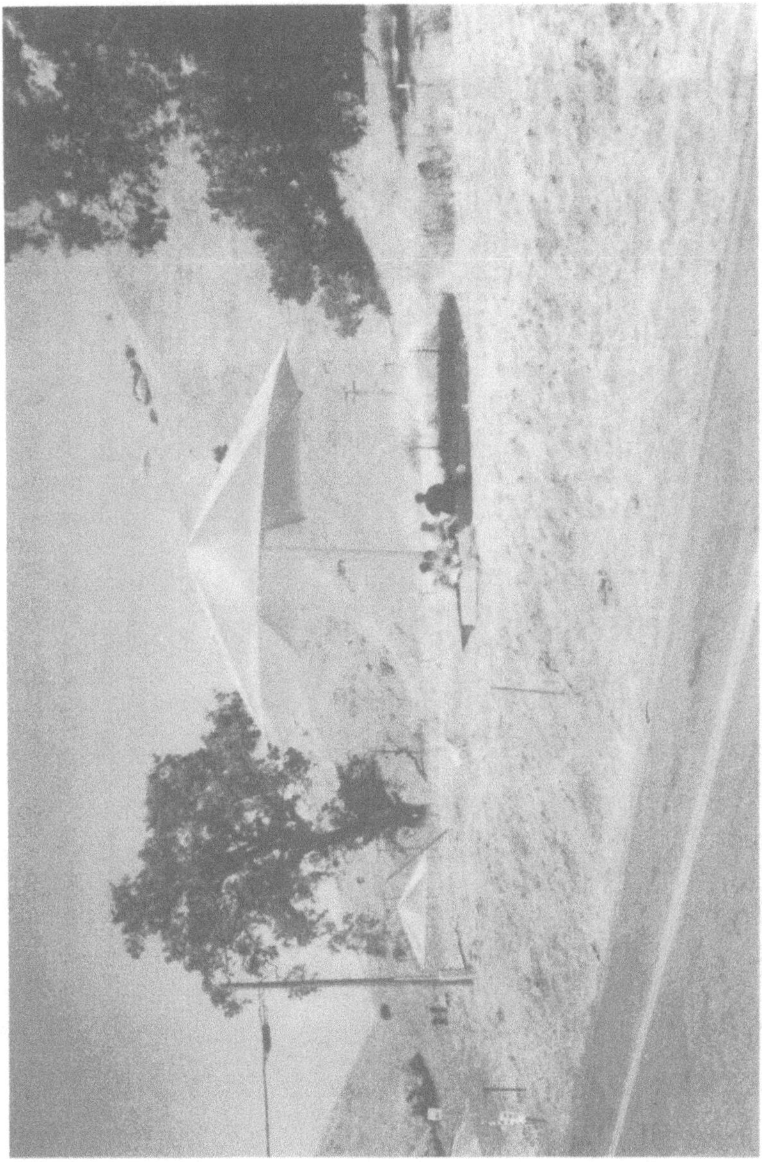

7. Christo. *The Umbrellas: Joint Project for Japan and U.S.A.*, 1990, the Tejon Pass, U.S.A.

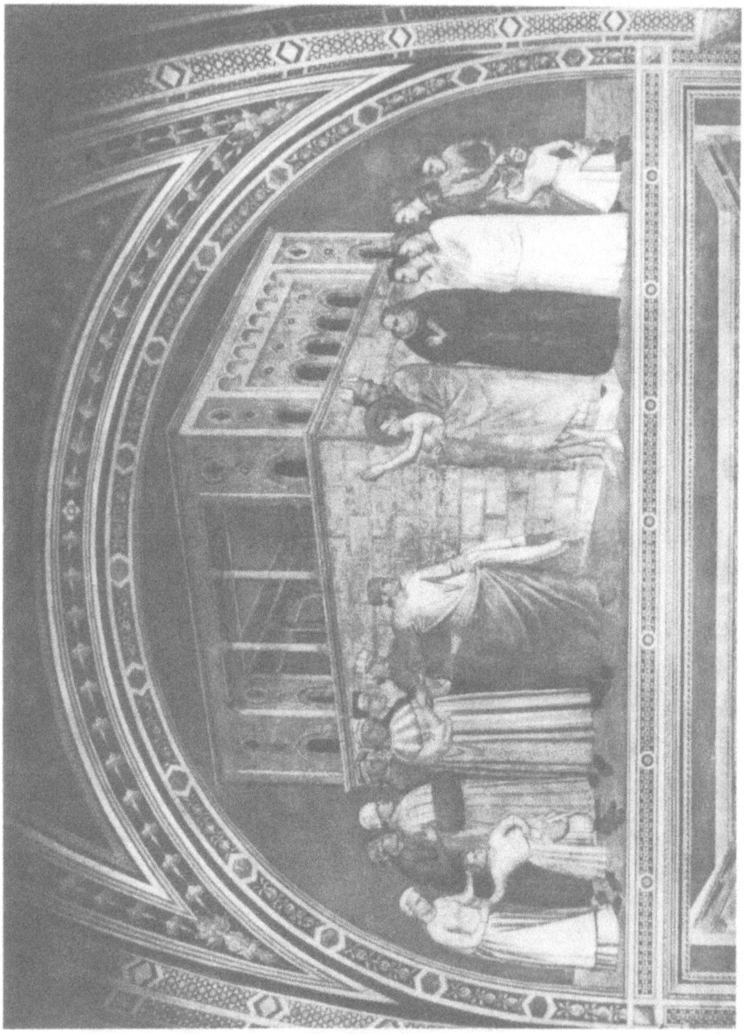

8. Giotto di Bondone. Scene from the Life of St. Francis fresco cycle: *St. Francis Renouncing his Father*, ca. 1269-1300. Bardi Chapel, S. Croce, Florence, Italy. Alinari/Art Resource, NY.

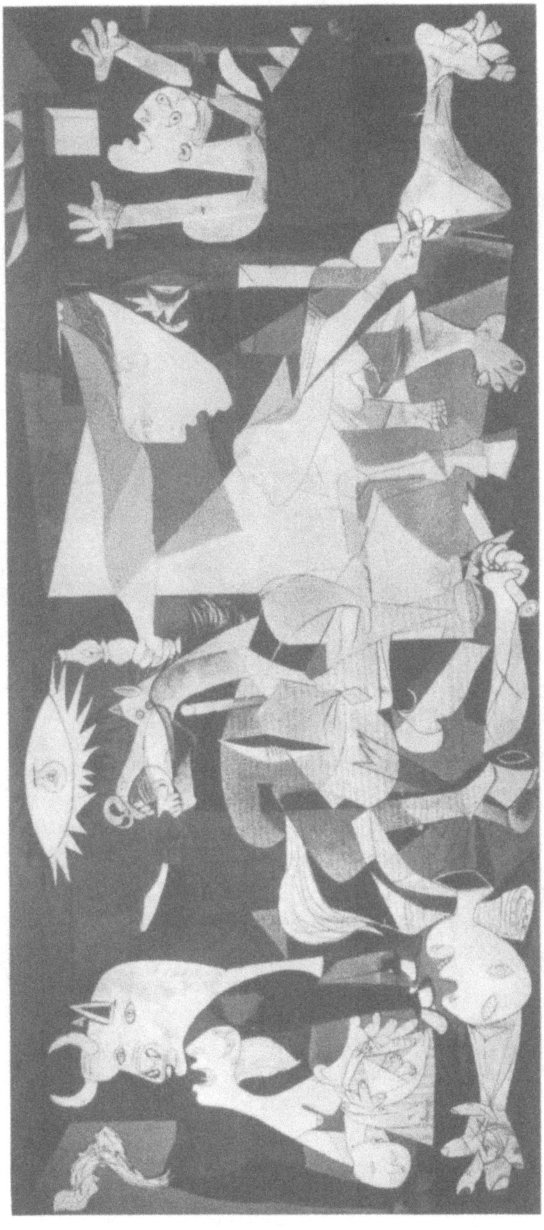

9. Pablo Picasso. *Guernica*. 1937, mural. (*Copyright ARS) Centro de Arte Reina Sofía, Madrid, Spain. Giraudon/Art Resource, NY.

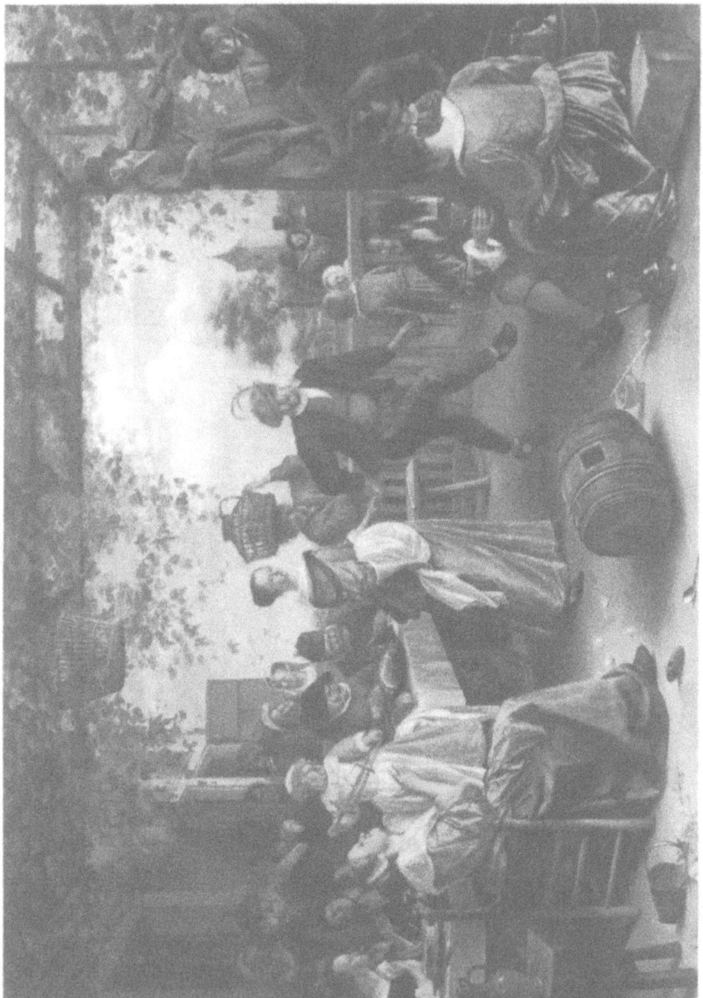

10. Jan Steen. *The Dancing Couple*, 1663, oil on canvas. The Widener Collection, © 1994 National Gallery of Art, Washington.

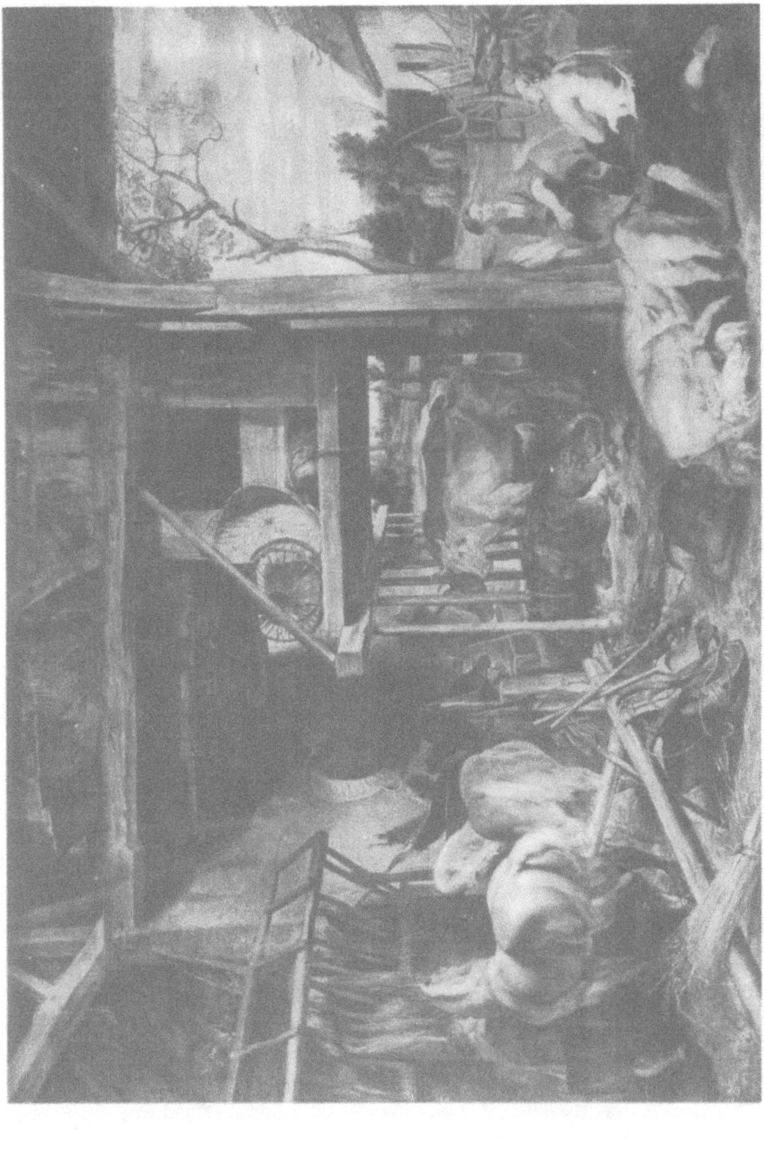

11. Peter Paul Reubens, (1577-1640). *The Prodigal Son.*, oil on canvas. Musée Royal des Beaux-Arts, Antwerp, Belgium. Giraudon/Art Resource, NY.

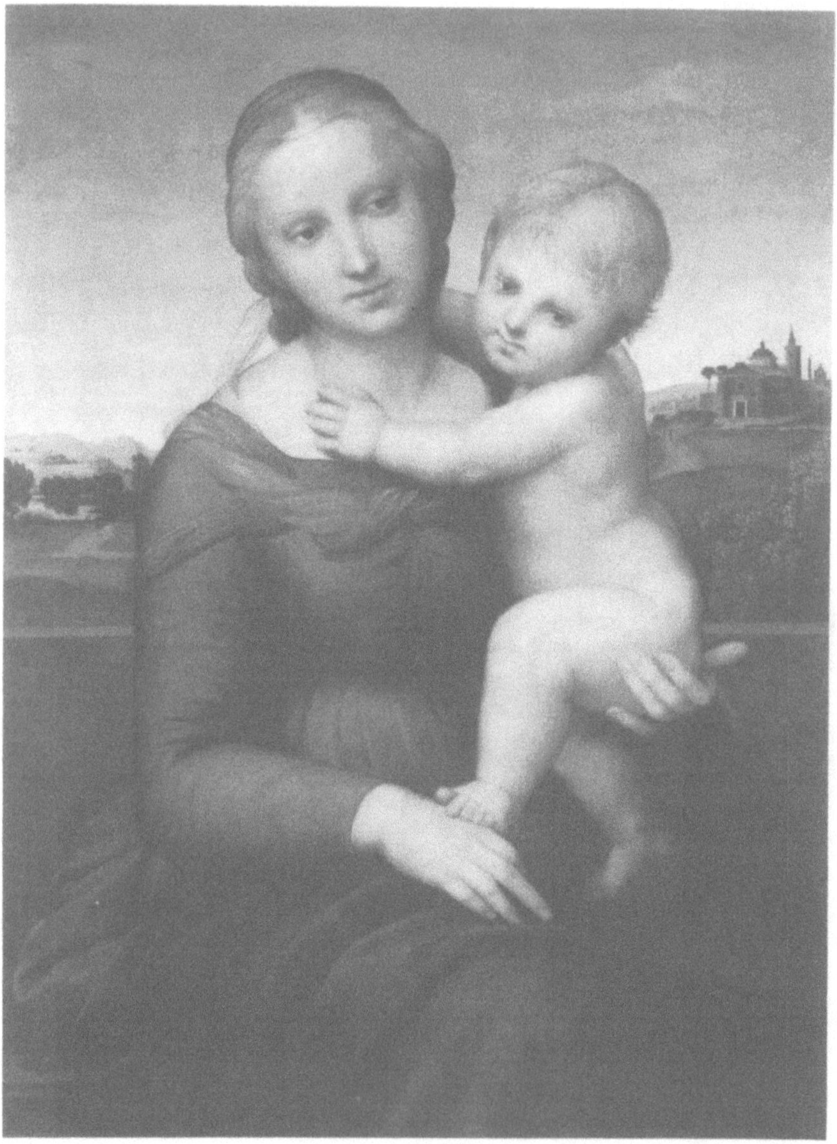

12. Raphael Sanzio, *The Small Cowper Madonna,* ca. 1505, oil on panel. The Widener Collection, © 1994 National Gallery of Art, Washington.

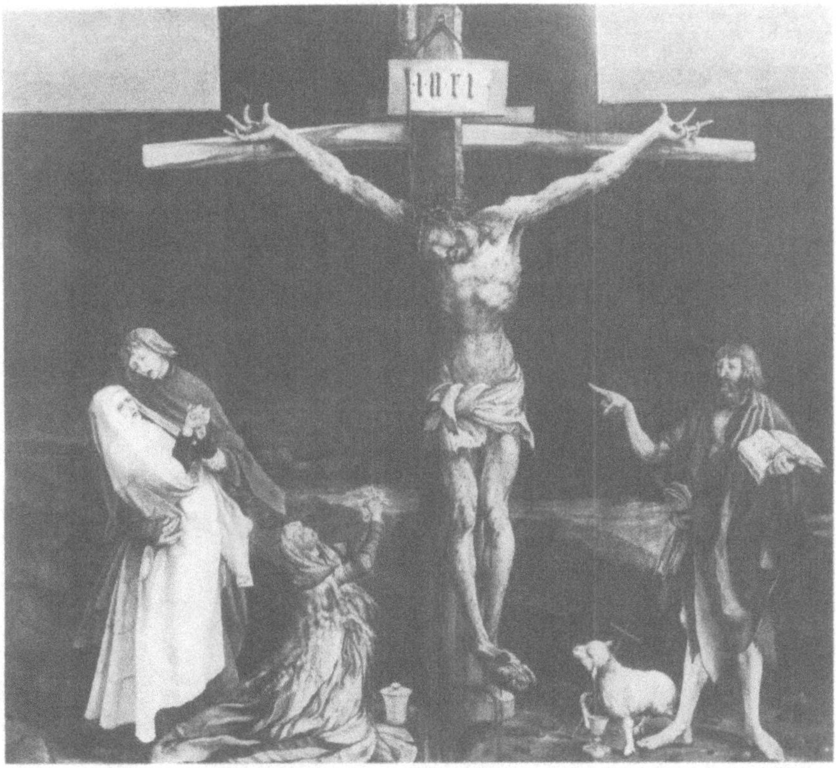

13. Mattias Grünewald, *Crucifixion from the Isenheim Altarpiece* (closed), 1515-1516, oil on panel. Musée Unterlinden, Colmar, France. Foto Marburg/Art Resource, NY.

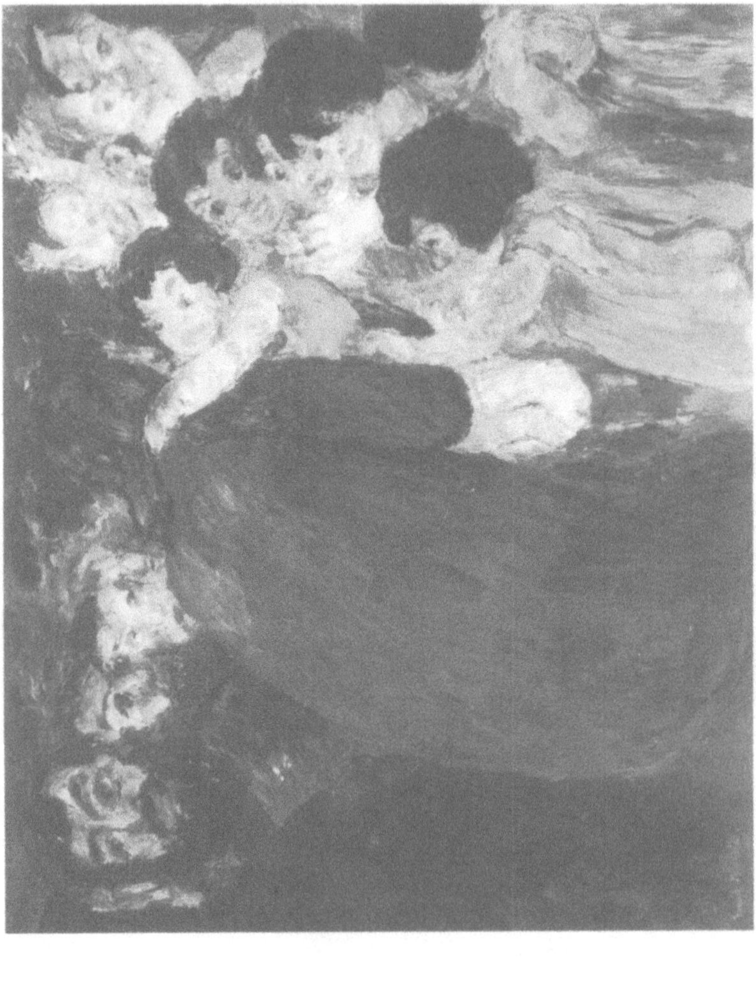

14. Emil Nolde. *Christ Among the Children*, 1910, oil on canvas, $34^{1}/_{8}$" x $41^{7}/_{8}$". The Museum of Modern Art, New York. Gift of Dr. W. R. Valentiner.

PART II:

PAUL TILLICH AND ART

About Paul Tillich

Paul Tillich (1886-1965) was a Protestant theologian contemporary with Polanyi. The son of a Lutheran pastor in Germany, his upbringing was austere and negligent of the visual arts. As a child, Tillich was interested in architecture and the world of vision about him, but he was encouraged to study music and literature. While serving as an army chaplain in World War I Tillich discovered the power of the visual arts. He found solace from the terrors of war in reproductions, poor as they were, of works of art. Yet not until he had the opportunity to engage with works of art directly while on a furlough in Berlin did he realize art's revelatory power.

The experience that proved crucial to Tillich's philosophical and aesthetic development was with Sandro Botticelli's painting, *Madonna and Child with Singing Angels* (Plate 5). Out of his experiences with visual art, Tillich developed a system of art categories for a theology of culture, namely, form and substance. Most important to Tillich's development of a philosophy on religion and culture was German Expressionism. Here is an artistic style which does not accept what life presents but breaks through it in order to reveal its depths.

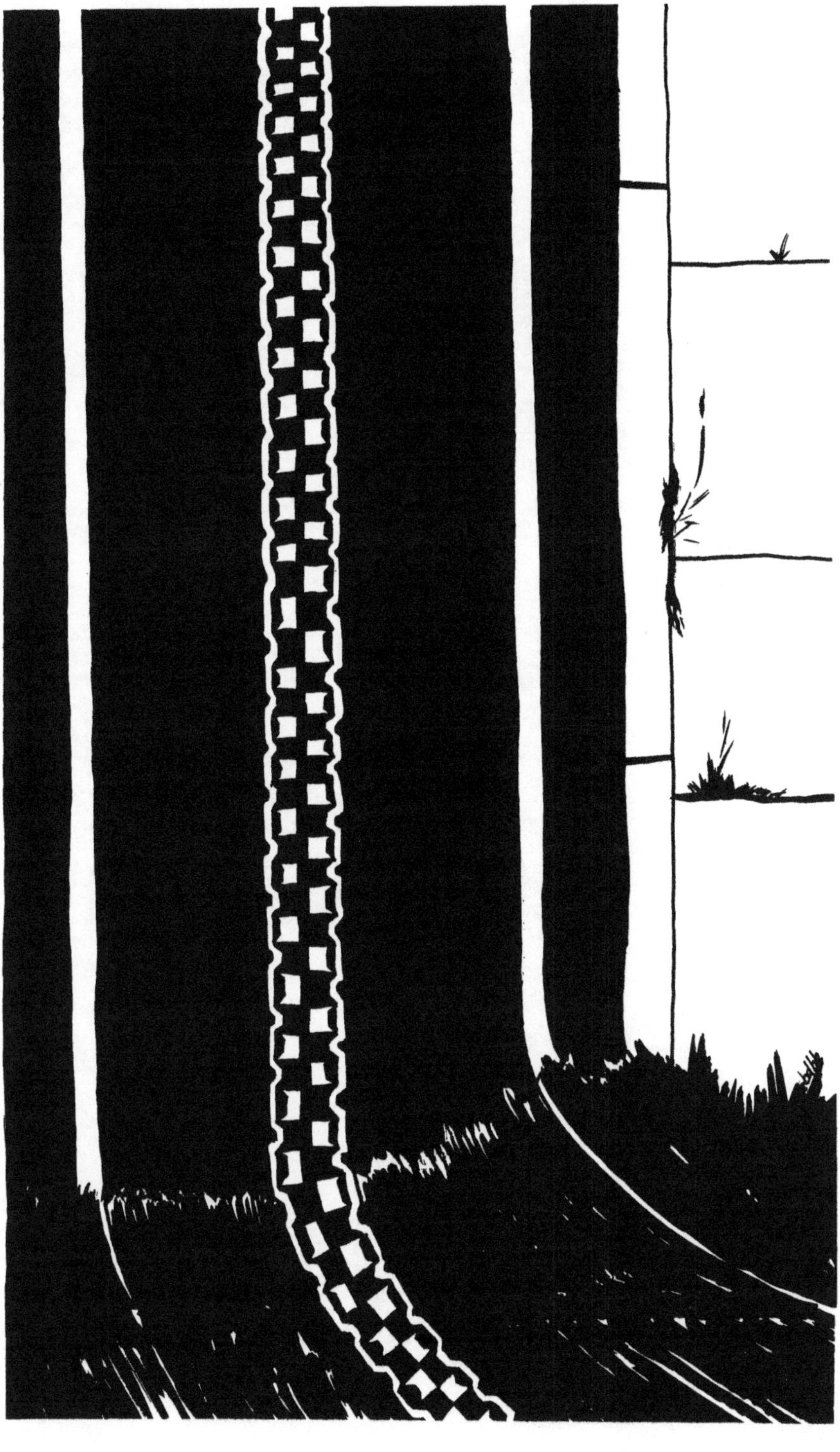

Chapter Seven

The Religious Element in Art

Polanyi found his ideas about theology to be confirmed by many of the passages in the writings of Tillich.[1] He is hesitant to assert that his method and scope agrees entirely with Tillich, but Polanyi does cite extensively examples of Tillich's writing that affirm his own thought. Tillich was a Protestant theologian who was interested in the arts. His writings on art have had a large influence on theologians and their acceptance of art as theological. Setting Polanyi aside, we enter Tillich's aesthetics to uncover his contribution to the field of art and religion.

Evidence of Tillich's system of categorizing art can be found in his 1952 lecture for the faculty, students, and friends of the Minneapolis School of Art at the Minneapolis Institute of Arts. In his third lecture, entitled "Religion and Art," Tillich describes four levels of relationship between religion and art. The first level is the general expressive power. Here Tillich maintains that all art expresses an ultimate concern through its style, and therefore, all art is essentially religious.

While all art is implicitly religious, the second level is delineated by being explicitly religious in style. The ground of being is disclosed. Religious content is the realm of the third level. Tillich combines the second and third levels for his fourth level, dealing with (1) religious style and (2) religious content; they both explicitly reveal ultimate reality.

A key element in Tillich's categorization of art is the notion of a content (his term for subject matter) that can be broken-through by an expressive power that frees the content from its surface reality so that its depth may be plumbed. Tillich's use of the term "content" to refer to subject matter rather than to the artwork's meaning is confusing. However, the shift into Tillich's definition of "content," is necessary to remain true to his thought. Understanding what is, for Tillich, the essence of the term "content," as well as "religious" and "style," is our present task.

In the first level where he states that all art is implicitly religious, Tillich is using a larger understanding of religion. For Tillich, religion is not an activity alongside other activities but an attitude of the spirit that points toward the Unconditional. Religion is not a special function of the human community, for all acts of humanity are pointed toward Ultimate Reality. Ultimate Reality grasps all reality, every function, every situation. "It is the depth of all cultural creations and places them in a vertical relation to their ultimate ground and aim."[2] Or as Tillich states in *Theology of Culture*, "religion is not a special function of man's [*sic*] spiritual life, but it is the dimension of depth in all of its functions."[3] Religion, then, is the depth of all cultural creations and functions. It is what "gives substance, ultimate meaning, judgement and creative courage to all functions of the human spirit."[4] Tillich's aesthetics proceeds from his theology of culture. He describes the function and meaning of art within this philosophy of religion.

In *What Is Religion?* Tillich outlines his theology of culture in an essay, originally written in 1919, entitled "On the Idea of a Theology of Culture." He begins by defining the cultural sciences which both perceive and create culture and the empirical sciences which only perceive reality. For the cultural as well as the empirical sciences, knowing is participatory. Humanity is estranged from reality, but in the pursuit of knowledge, the knower moves toward reunion with reality.

In the cultural and empirical sciences, the knower is caught within the tension between union with reality and reality's quality of "otherness." In the case of empirical science, the knower is detached from the object of inquiry. Empiricism is a "controlling knowledge" of certainty verifiable through repeatable experiments. Since reality already *is*, the empirical scientist does not create reality, but only observes and discovers its being. Our particular standpoint must be overcome so that reality can be discerned as "the criterion by which what is right is measured."[5] Only one reality exists, one right; all other judgments are wrong.

The case is different in the cultural sciences. Tillich recognizes the active participation of the knower and the role his/her intellectual framework plays. Culture is perceived and created out of the knower's personal conceptions. Matters of right and wrong lose their relevancy in his view since the perspectives toward reality are limitless.

Both the arts and religious studies are cultural sciences. Neither can be learned through abstract reasoning, since they are determined by individual standpoints creatively acting within an historical and cultural community. The universal forms arise from the individual's creative relation to culture and history. Here, the knower takes the object of reality into itself in a receiving type of knowing. Knowledge is verified through the experience of a creative union of the self with reality.

According to Tillich, certainty is operative in the empirical sciences, while significance is key to the cultural sciences. The cultural sciences, which add the task of creating, as well as observation and discovery, have three forms. The first form, the philosophical, is the arena of ideas. Whether it be the philosophy of art, culture, or religion, this form explicates the universal forms. Its thought is comprehensive. For example, the philosophy of art perceives the value and essence of art, unveils the universal forms and concepts of art, addresses the phenomenology of art. Tillich's statement that "all art is religious" would be appropriate within the philosophy of art. The philosophy of culture distinguishes the value and universality of cultural forms. Philosophy of religion informs us that all acts of humanity are pointed toward Ultimate Reality.

The normative sciences—aesthetics, ethnology, theology—are the second form of cultural science which provide a distinct systematic expression. Aesthetics, then, is not a branch of philosophy, but a normative science. Aesthetics, for Tillich, cannot be defined as the philosophy of art, for it is a science of art. Aesthetics and the philosophy of art are two forms, two poles on a line within the larger field of cultural science. Philosophy of art is an ontological pursuit that asks what art is, while aesthetics is existential because it questions what art does.

The philosophy of art looks at art's value and essence, while aesthetics, as science, asks, what is beauty? It must answer from a concrete standpoint. Beauty is discovered not as a universal idea but through the particular. While philosophy is a comprehensive questioning of the nature of art and its import, aesthetics is the concrete and creative realization of the norms of art. Aesthetics is the content or systematic expression of what art does. For Tillich, this would be the relation between form or the surface reality of art and its import or depth-content.

The same could be said for theology, which is a science of religion, and for ethnology in correspondence with the philosophy of culture. The forms of philosophy and the normative sciences are inseparable. Each exercises an influence over the other. Theology provides the concrete basis to the universal pole of the philosophy of religion; without this concrete basis, philosophy is useless. Theology and aesthetics give scientific significance by way of the particular to the philosophies of religion and art.

Tillich's categorizations of art are a dialectic play between art's form and its import or expressiveness, which is evinced through art's style. Tillich does not separate import and form but finds them to be essentially didactic. Tillich's discussion of style and the categories that specify an art's religiosity are part of his philosophy, while his discussion on the relation of content to import would be a part of his aesthetics. This interplay is manifested and the particular to which these normative sciences refer is answered in the third form of cultural science, the philosophy of history of cultural values.

The history of cultural values is the means, the mediator, between the scientific and philosophical poles. Though history, theology and aesthetics give objective scientific significance to philosophy; it validates theology and aesthetics. The history of cultural values provides the transition from the universal to the particular, and from the particular to the universal. As a form, it classifies while at the same time it is the means for both philosophy and normative science. Neither pole–aesthetics and the philosophy of art–are detached from actuality but both find their authenticity in history. Because actual people actually create art, the philosophy of art and aesthetics arise. Because actual people worship and practice religion, the philosophy of religion and theology arise.

While art, culture, and religion, have the same forms within cultural science, their likeness disappears when we consider cultural function. Whereas the two poles within cultural science are philosophy and normative science, the two poles within cultural function are (1) the subjective, theoretical pole where, Tillich states, the "spirit absorbs the object into itself," and (2) the objective, practical pole in which the "spirit penetrates the object and molds it after itself."[6] Within the first pole lies the intellectual and aesthetic pursuits where the mind receives reality objectively. In the second pole lies the individual and socio-ethical pursuits, such as law and society, in which the mind injects itself into reality in order to transform reality into what the mind deems reality should be. Neither art nor religion can be placed within the theoretical or the practical spheres of culture. Art is a cultural creation not a function; religion is a belief that transcends both of these spheres.

Art is neither a purely aesthetic experience nor a function of culture. As James Luther Adams explains:

> If [art] were theoretical, its aim would be that of *Theoria*, the pure apprehension of the essence of things. If it were practical, its aim would be to represent ethical ideals in tangible form. The touchstone of art is its force of expression, not the ideal nature of what is expressed.[7]

Robert P. Scharlemann, however, places art within the theoretical sphere of Tillich's system. Scharlemann understands Tillich to assert that art serves the function of receiving reality rather than that of transforming reality.

> Despite the fact that the making of a work of art might seem to involve a doing, or a transformation of reality—a sculptor does seem to transform a boulder into a statue—the relation of self to world that is implied in the work is still theoretic. Its intention is not to make reality different from what it is but to show a dimension of reality that does not otherwise appear.[8]

Art implies a relation in which the mind conforms with the real. But rather than focusing on the content or surface reality, as does cognition, art expresses the import or depth-content of reality. Cognition grasps the forms of reality in order to recognize it; art expresses the import in order to interpret it. Yet in both cases the mind conforms itself to reality, and reality is not itself transformed. Thus, Scharlemann defines art according to Tillich as, "an object that is the product of a theoretic relation between the self and the world whose intention is to express the realness of the real in the object rather than to grasp its form."[9]

The key element in the placement of art as *Theoria* or as a cultural creation that stands outside of or borders the functional sphere is expression. For Scharlemann, expression is a theoretical activity of interpretation; for Adams, expression is more than *Theoria*, since expression is not bound to its content in order to achieve import. Content may lose its relation to form; it is not essential to the task of artistic expression.

In Tillich's thought, expression is the first characteristic of artistic creativity. The artist expresses a quality of reality that transcends mere objectivity and mere subjectivity. Art overcomes and transcends the subject-object split, for it expresses the ultimate love union of the two. Expression is driven by love, the human urge to conquer estrangement through achieving reunion with the separateness of all being. Art does this by reuniting us with the meaning in all aspects of reality as it expresses the depths of things that are beyond the reach of cognition. Art functions to unify the epistemological polarities in which the knower is caught as s/he pursues reality, as do the cultural sciences in general. In this sense, Scharlemann is correct in his evaluation.

However, Tillich states that the second characteristic of artistic creativity is transformation. Art transforms given reality into a form which expresses the aesthetic dimension of reality. The visual form of the visual arts is never the same as the experiential form that can be known through any and all of the senses. Reality is transformed so that it becomes a symbol for the depth-content of that given reality. Herein lies

Tillich's definition of beauty. Beauty is the power of mediating import through transforming experiential reality into artistic form. Creativity, then, is a revelatory experience as the artist is grasped by the joy of love that drives toward reunion through the artist's creative encounter with the ground of being that permeates and gives meaning to everything. The artist expresses the "courage to be," the experience of the ground of being that transcends all of life and its ambiguities, in artistic form.

The artist penetrates surface reality in order to reveal transcendent reality. The artist participates in and creatively responds to life and transforms it so that we may see anew. Art presents the ambiguities of life as they are united in the love that transcends their polarities. The polarities of life are not dissolved but held in a dynamic tension of balance. Art thus presents and participates in Tillich's symbol of the Kingdom of God where estrangement is conquered and the polarities of life are united (not dissolved) in love, in Being itself.

We might say that art creates a "virtual Kingdom of God," for it anticipates salvation. And anticipation is the third characteristic of artistic creativity. Both the ideal beauty of perfection and the expression of imperfection create a nonclassical perfection that anticipates the Kingdom of God or salvation. Therefore, art transforms since it also expresses and anticipates. Reality is transformed in its presentation, not in its actuality. Though our experience of reality will be changed when we encounter art. Art in its transformation does not control reality but penetrates its depths. Art is neither theoretical nor practical but transcendent; it resides on the border as Adams maintains.

Scharlemann critiques Tillich for not taking into account the difference between aesthetic and cognitive forms. Cognitive forms, say, in the case of language, can be paraphrased, translated, or stated in more than one way, while sensuous (aesthetic) forms cannot. Sensuous forms elude cognitive expression. The forms of art are unique and non-transferable. This difference between cognitive and aesthetic forms is not between grasping forms and expressing depth; it is in the forms themselves. Tillich does understand the uniqueness of aesthetic forms when he recognizes that form is not bound by content. If we view art as a theoretical cultural function, we, like Scharlemann, cannot appreciate Tillich's sensitivity to the uniqueness of aesthetic form. If we agree with Adams and place art as a cultural creation on the border of cultural functions, then we will recognize that Tillich did take into account the difference between aesthetic and cognitive forms.

Religion is not a cultural function. It is a principle actualized in cultural functions. Religion, as such, does not exist as a separate sphere; it is not a thing nor a function, but an attitude or belief. No special religious sphere exists; religion is always known through culture. Therefore, no real conflict occurs between church and state because the two spheres do not exist separately. Likewise, no conflict occurs between sacred and profane art, again because the two separate spheres do not exist.

In "On the Idea of a Theology of Culture," Tillich defines religion. The passage is lengthy but pivotal to understanding Tillich and his aesthetics.

> Religion is directedness toward the Unconditional. Through existing realities, through values, through personal life, the meaning of unconditional reality becomes evident; before which every particular thing and the totality of all particulars—before which every value and every system of values—before which personality and community are shattered in their own self-sufficient being and value. This is not a new reality, alongside or above other things: that would only be a thing of a higher order which would again fall under the No. On the contrary, it is precisely through things that reality is thrust upon us which is at one and the same time the No and Yes to every thing. It is not being, nor is it the substance or totality of beings; it is--to use a mystical formula—that which is above all beings which at the same time is the absolute Nothing and the absolute Something. But even the predicate "is" already disguises the facts of the case, since we are here dealing not with a reality of existence, but with a reality of meaning, and that indeed is the ultimate and deepest meaning—reality which shakes the foundation of all things and builds them up anew.[10]

Thus, Tillich can state that all art is essentially religious. He has negated the possibility of a separate religious sphere. Religion is a directedness toward the Unconditional; it is not a being, a substance, or a totality of beings. It is not a reality of existence but a reality of meaning. We give art its value as art, not because it exists but because it is meaningful.

What about the separate sphere of religion that does exist? Church and state are in conflict. The existence of a separate sphere of religion, institutional and personal, is due to humanity's tragic estrangement of the spiritual life from its own ground and depth. In *Theology of Culture*, Tillich explains that religion opens up the depth of our spiritual life. It enables us to move beyond "the dust of our daily life and the noise of our secular work" to experience Ultimate Reality. However, religion errs by making itself, its doctrines, rites, and laws, ultimate. It lords over the secular realm and persecutes dissenters. "It forgets that its own existence is a result of man's [*sic*] tragic estrangement from his [*sic*] true being. It forgets its own emergency character."[11]

Religion and the secular realm are separated due to estrangement. In estrangement, humanity turns away from the Holy and elevates itself as the supreme center of the world. Through hubris we view the world emanating from ourselves. This estrangement drives life toward one or another of the poles of ambiguity, disrupting the essential unity of life and the unity of truth. We are driven to separate the sacred and the profane in order to reflect upon them and be aware of them. But if in reflection we can recognize the common experience of ultimate reality in both the sacred and profane spheres, then the conflict between the religious and secular is overcome. This overcoming of the religious-secular conflict through the realization of their mutual immanence in which "religion is the substance of culture, and culture the form of religion,"[12] epitomizes Tillich's theology of culture.

For Tillich no distinctly religious and secular spheres exist. Science and other cultural realms are protected in their autonomy. Religion cannot exercise heteronomy. Instead, science, culture, and 'religion' are united under the paradoxical love that threats and supports. The cultural spheres while remaining autonomous are not separate from religion, but, like religion, they are directed toward ultimate reality of meaning in their depths. How this is so is evident in the basic categories of Tillich's theology of culture which is derived from his study of art.

Like art, theology of culture can be categorized by its content, form, and import. Content, in Tillich's scheme, is simply the objective existence of subject matter given form. It is drawn from the inexhaustible multitude of objects encountered by the human mind in sensory images. Every encountered reality can potentially become artistic content and is not limited by qualities such as good or bad, beautiful or ugly, whole or broken, human or inhuman, divine or demonic.[13] The selection of content is dependent on form and import and is particular to a given culture and historical period. Content varies from one artist and his/her artistic period to another.

Form is what makes a thing what it is; it is the artifact's materiality and being. It transforms content into a work of art through an expressive use of particular materials, such as color, canvas, stone. Form is the ontologically decisive element in art.[14]

Import is the substance that gives the form and content its significance and its validity. This is the essential religious element. While through form content is made into an artistic creation, import refers to meaning. This meaning, claims Tillich, is always imbued with spiritual significance. We grasp meaning in a work of art through the formal expression of content. "Content is accidental, substance essential, and form is the mediating element."[15]

We have discovered the meaning of the first two categories of Tillich's theology of culture in his 1957 article, "Protestantism and Artistic Style." He describes three elements common to every work of art: content, form, and style. His theology of culture consists of the three elements: content, form, and import. Style is a special quality of the form through which import is expressed. Style is particular to a history and a culture, and to Tillich, the second level of the relation between religion and art.

In Tillich's theology of culture, the autonomy of cultural functions is grounded in form, and the theonomy, where culture and religion meet, is grounded in import. In other words, culture is meaningful in its expression of form. Form and content are not opposites but stand at one pole, while import stands at the other pole (Figure 2). Content is the least significant and may even lose its relation to form with an abundance of import. While content may lose its relation to form, form and import are always in a necessary relationship. Form cannot exist without import lest it fall into a shallow autonomy in which law is empty and meaningless; import cannot exist without form lest it fall into the worst kind of heteronomy in which the import, which is only a partial aspect of the unconditioned validity of the divine, claims absolute authority over all other reality.

Form and import are two poles on a line that are always mutually immanent. While the relationship always exists between form and import, a predominance of form can occur, and thus, greater autonomy, or a predominance of import, in which form becomes inadequate and is shattered while maintaining itself as form. When form is shattered greater theonomy is presented. Form is imbued with the Unconditioned so much so that it cannot contain it; form is broken-through and shattered, content is detached, and the existing shattered form is in immediate relation to the import.

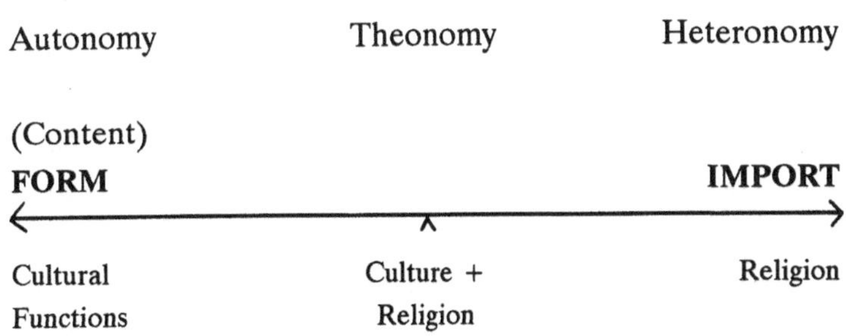

Figure 2. The poles of form, content, and import.

Cultural functions may exist in form without import. Importless form is an autonomous state in which the function is an empty shell and meaningless. Form can exist without import. Import, on the other hand, is dependent on the existence of form. Without form, nothing exists for import to transcend. Thus, if import existed we would not know it because no separate religious sphere exists. We can only recognize the Unconditional through form. In the case of heteronomy, a single form is grasped, and this one form is mistaken as the absolute, but in reality the absolute transcends this form and all other forms as theonomy.

Art, in order to be art by Tillich's definition, must have form and import. The form causes the art to exist, as in the case of cultural functions. An art that is merely meaningful without a connection to our experience of reality, or form, may be "beautifying," but it does not fulfill aesthetic function. Likewise, naturalistic art that does not convey import may be illustrative of reality, but does not fulfill its aesthetic function. Art must also have import. All art is religious, therefore, all art has import. Autonomous art, art without meaning, is not, for Tillich, art. It is perhaps illustration or craft. Art without import would be the worst kind of kitsch, not art.

Think of the anti-art Dada movement in the early 1920s, particularly Marcel Duchamp's Ready-mades. If Tillich were asked if the *Bicycle-Wheel* were art, his response would be: Is it meaningful? Does it reveal ultimate reality? If yes, then it is art; if no, then it is not. Art is different from other cultural forms in that it must have import, it must be theonomous, it must be essentially religious. Tillich in his utopian view of a theology of culture would want to assert that all cultural forms are theonomous. However, at realistic moments, especially later in his life, he recognized that this is not the case. Autonomy and heteronomy will continue to manifest themselves in culture.

Art, a cultural creation, is on the border between cultural functions and religion. It must partake in both to be art. If we recall our definition of estrangement in which we are driven toward one pole or the other, we could say that art, if only partially, conquers estrangement because it is essential for art to evidence both poles. No heteronomous nor autonomous art exists, only theonomous. Art may lean toward one of these poles or the other, but not completely so.

The task of a theology of culture is to give expression to the form-import relation, though not from the standpoint of form, for this is the task of aesthetics and art criticism (in the case of art), but by taking import as the starting point. Here is the sphere of theology of art as well as theology of culture. Neither of these are the systemization inherent in the normative sciences, theology, and aesthetics. The theology of culture (and art) stand beside the normative sciences in order to analyze cultural creations from a religious perspective. From a religious standpoint, cultural creations are classified according to type and historical-philosophy, and an ideal outline of a religious culture is produced. In Tillich's words, a theology of culture

> has a threefold task, corresponding to the threefold character of the cultural-systematic sciences in general and the systematic science of religion in particular:
>
> 1. General religious analysis of culture
> 2. Religious typology and philosophy of cultural history
> 3. Concrete religious systematization of culture[16]

In the first task, the relation between form and import is discerned. In this discernment of import as expressed through form the religious reality appears with its Yes and No to all things. The theologian of culture is required to ascertain the relation of the Yes and the No, and the force in which both find expression. The possibilities of expression are infinite within these relationships.

The first task of an art theology would be to perceive the relation between form and import from the standpoint of import, and then to discover the concern about being or not being at all, the force of the Yes and the No as expressed in the artwork. Some questions to consider might include: Is the Unconditional a threat or a support? Is a freedom to deny or affirm existence present? Is being conveyed through dread or hope? Is the force of denial or affirmation stronger? Is the form destructed or preserved?

In the second task, the theologian of culture is called to construct a religious typology and philosophy of cultural history. The concern here is the various relations of form and import on the line that divides the two poles; the possibilities are infinite with many intermediary stages and transitions. Tillich defines the three fundamental types based on the three decisive points of the line: the two poles and the center in which form and import are in equilibrium.

First, when form is predominant, the secular or formal cultural creation eclipses import; on the second, or import-dominated, pole is the religious-cultural creation; third, in the center is the balanced classical-cultural creation (Figure 3). When the theologian of culture relates these typologies to the present and past systematically, a historical-philosophical classification develops.

The art theologian's second task is to construct a religious typology for art and then develop a philosophy of art history. Following Tillich's three fundamental types of cultural creations three typologies of art can be cited: (1) the form-dominated type which includes impressionism and realism; (2) the import or *Gehalt*-dominated romanticism and expressionism; and (3) the balanced idealism and classicism (Figure 4).[17]

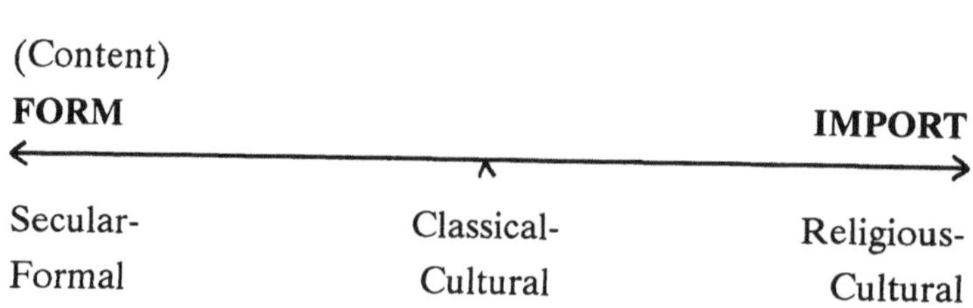

Figure 3. The Poles of a Religious Typology

Each of these types, which are derived from the three decisive points on the form-import pole, contain two stylistic types, which reflect the difference between subjectivity and objectivity. Remember that art transcends subjective-objective experience and is never wholly subjective or objective in Tillich's categories. In the form-dominated type, impressionism is dominated by the subjective pole because it breaks the surface and shatters form, while realism, with a preponderance of objectivity, clarifies the surface as it drives toward a sharpening of form. These two types can be said to be non-religious in the narrow sense of religion. Even when treating religious subject matter, a depth-content of any substance is absent; this secularization is only relative and not absolute. Tillich cites social scenes of extremely secular character as exemplifying art that is implicitly religious though not explicitly so in subject matter or style. Jan Steen's *The Dancing Couple* (1663) (Plate 10) and Henri Toulouse-Lautrec's *A Corner of the Moulin de la Galette* (1892) are cited as examples of a form-dominated type of art: Steen on the objective pole of realism, and Toulouse-Lautrec on the subjective pole of impressionism. Yet both, while thoroughly secular, express the power of being and thereby are indirectly religious.[18]

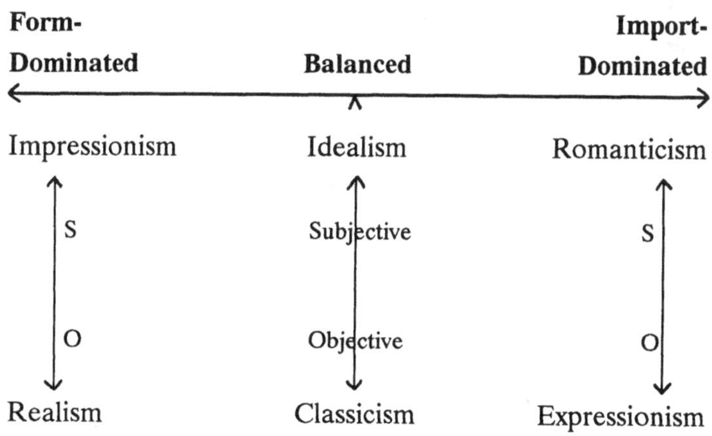

Figure 4. Tillich's Religious Typology of Art

Import-dominated romanticism or concrete expressionism breaks through pictorial form with its preponderance of subjective feeling. Import-dominated art on the subjective pole, such as the Baroque, is "revolutionary and of ecstatic piety with the coloring of aristocratic spirit."[19] Abstract expressionism[20] is concerned with an objective spirituality. "It is a form of revelation for an objectively valid, mystically immediate spiritual substance of an era or cultural circle: it is the hieratic—and perhaps, one day, the socialist—style."[21] In these two types a depth-content of the world is explicitly revealed as it breaks through form. A still life by Cézanne or an animal painting of Marc can become as "sacred" as a given religious content. While expressionism—concrete or abstract—is, for Tillich, overtly religious in style, it, too, can be secularized by an artist who paints in the style without a religious disposition or an experience of the style's depth-content.

Within the balanced-type, idealism is subjective because it intensifies the universality of an individual type and delineates life potentialities. Classicism, on the objective pole, focuses on the perfection and completeness of real form.

The third task of the culture theologian is to approach culture from the standpoint of import to outline a religious culture system. The task is not to create a religious culture but to recognize the all-embracing unity of cultural functions. S/he demonstrates that import transcends and breaks-through all the different phenomena of culture. The theology of culture may move beyond the actual expressions of import to a visualization of a truly religious system of culture. In so doing, the theology of culture does not create the religious system of culture but helps unify the existing form of culture through import.

The development of a religious system for art is the third task of the art theologian. Again, the art theologian does not create religious art but identifies art as religious. The art theologian views the form of art from the standpoint of import not to identify religious subject matter but to discern the religious expression in all possibilities of content and form. As a result of this discernment, the conflict between art and systematic forms of cultus is bridged. The vision of art, religious in essence, is developed. Tillich's Protestant Principle is operative here: "God is present in secular existence as much as he [*sic*] is present in sacred existence."

In sum, the threefold task of a theology of art (1) establishes the relation between content, form, and import; (2) constructs a systematic typology of artistic styles based on the form-import relation found in works of art; and (3) visualizes an authentic religious artistic style that unifies the depth-content presented in all variations of aesthetic form.

A theology of art recognizes that the holy can be grasped through artistic form. The form is not holy, but the Unconditional that is mediated by and through the form is holy. From the standpoint of import, the art theologian discerns the import or meaning that is breaking-through and being expressed through form. And all art to some degree expresses the mediation of the Unconditional. In this sense all art is religious.

Tillich outlined a cultural-theological consideration of Expressionism as exemplary of his theology of culture. His choice to analyze Expressionism is due to his conviction that Expressionism is an impressive example of the relation between form and import, and because this school influenced his whole development of a theology of culture.

In Expressionism, content, or subject matter, has lost its significance. External reality has been broken-through so that its depths may be visible. The horror that lies within is made visible and is deepened by the cosmic sense of the guilt of sheer existence. Form is shattered by a religious import seeking to find form. Expressionism is the paradox of the struggle made visible between theonomy and autonomy. One pole of this visible paradox is the horror that dwells within all living creatures; the other pole is a mystical love that dissolves individual distinction so that all living creatures may be unified, and thus, redeemed. The form-destroying No predominates within a profoundly expressed tension between the No and a new and absolute Yes. The Yes is the weaker element since the artists intend a strong religious passion of Yes yet are overwhelmed by the suffering No.

Performing the task of a theologian of culture, Tillich has discriminated the relation between form and import (Expressionism is an import-dominated type) and the force of the No and the Yes (the No predominates). His description of Expressionism is general, for he does not cite examples for clarification or support of his ideas. How exactly did he discern the manifestation of the No and the Yes in the paintings of Expressionism? How is Expressionism a more theonomous school than, say, one of the balance-types such as Greek classicism? An investigation of the character of an expressionistic style follows to elucidate an answer to these questions.

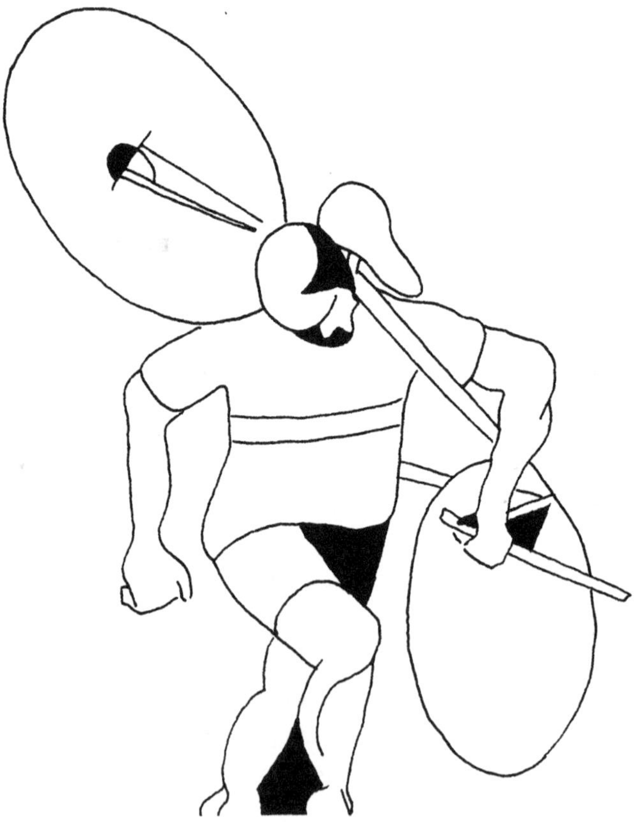

NOTES

1. Michael Polanyi, *PK*, p. 283, note.
2. Paul Tillich, *Systematic Theology*, Vol. 3 (Chicago: University of Chicago Press, 1963), p. 158.
3. Paul Tillich, *Theology of Culture* (*TC*), ed. Robert Kimball (New York: Oxford University Press, 1959), p. 6.
4. Ibid., p. 9.
5. Paul Tillich, "On the Idea of a Theology of Culture," in *What Is Religion?* (*WR*), ed. James Luther Adams (New York: Harper and Row, 1969), p. 155.
6. Ibid., p. 161.
7. James Luther Adams, *Paul Tillich's Philosophy of Culture, Science, and Religion* (New York: Harper and Row, 1965), p. 67.
8. Robert P. Scharlemann, *The Thought of Paul Tillich, An American Academy of Arts and Science Book*, James Luther Adams, Wilhelm Pauck, and Robert Lincoln Shinn, eds. (San Francisco: Harper and Row, 1985) p. 161.
9. Ibid., p. 160.
10. Tillich, "On the Idea of a Theology of Culture," p. 163.
11. Tillich, *TC*, p. 9.
12. Paul Tillich, *The Protestant Era* (Chicago: University of Chicago Press, 1948), p. 55-7.
13. Paul Tillich, "Protestantism and Artistic Style," (1957), in *OAA*, p. 120.
14. Ibid., p. 120.
15. Tillich, "On the Idea of a Theology of Culture," p. 165.
16. Ibid.

17. This discussion is based on Tillich's 1921 article, "Religious Style and Religious Material in the Fine Arts," written two years after "On the Idea of a Theology of Culture." The article can be found in *OAA*, pp. 45-57.

18. Tillich, "Art and Society" (1952) in *OAA*, p. 35.

19. Tillich, "Religious Style and Religious Material in the Fine Arts" (1921) in *OAA*, p. 54.

20. Ibid., p. 53, note 5. John and Jane Dillenberger point out that by abstract expressionism here, Tillich is referring to the early abstractionist, such as Kandinsky and not to the 1950s Abstract Expressionists of New York.

21. Ibid., p. 54.

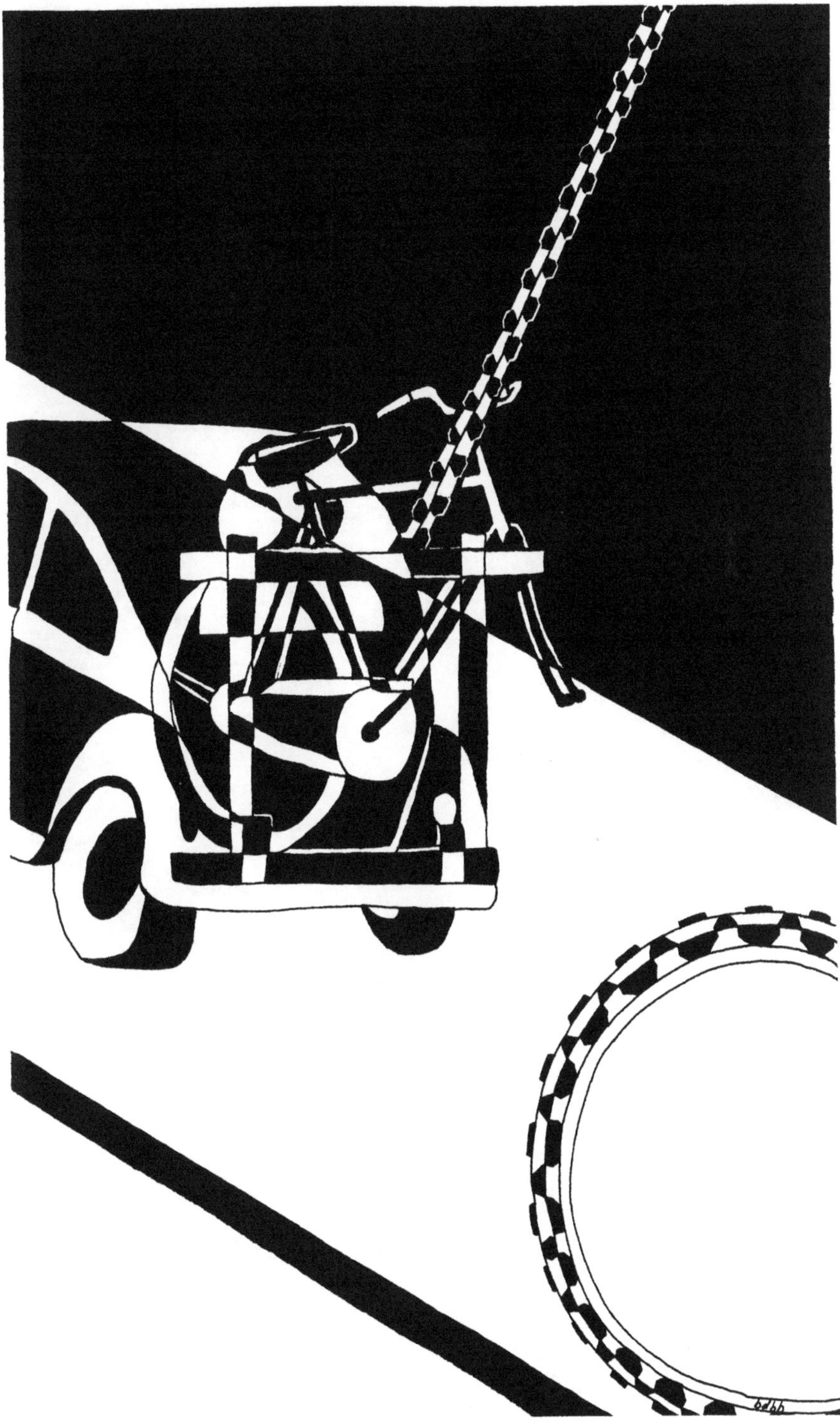

Chapter Eight

The Expressionistic Style

The second-level relationship between religion and art is art which points directly toward the ground of being through an expressionistic style where material from natural reality is used to symbolize ultimate concern.

First, let me establish a working definition of style. Style is a difficult concept to define as it is a mysterious cohesion of artistic expression that eludes discursive description. Like art, which cannot be fully defined discursively, style can only be known through the experience of the artwork itself. We can only know style tacitly as we attend to the artwork. In general, we could state that style is a given set of common characteristics that identify a work of art with a particular artist, school, movement, period, or geographical region. Style can be described through elements of form, which in painting may include brushwork, color, choice of motifs or iconography, line, mass, space, and light. While these characteristics may be identified with a school or period, they will also always be particular to a given personality. Style may be said to be the manner in which an individual expresses various sensory images through form from a given socio-historical location.

While Tillich understands art as what expresses the meaning or import of life, style relates that meaning to a history or culture. It is the element through which the experience of ultimate meaning and ultimate being are expressed. Style guides the artist's choice of subject matter and formal means of content. Because style mediates artistic expression it documents an epoch's religious sensibility. Art, through style, is attuned to the "situation of its time" since the artist creatively acts within an historical and cultural community.

Style reveals what cannot be grasped in any other way: the human encounter with reality in all its dimensions. This encounter cannot be expressed directly, but through style, the impression of the meaning of existence experienced by a culture is disclosed. We can discern the meaning of style through four approaches: (1) a general inquiry into the historical conditions of a group or individual; (2) a comparison of the style of one cultural creation with another from the same group, such as painting and philosophy; (3) an analysis of the subject matter chosen; and (4) an interpretation of the characteristic elements which constitute style in contrast to other stylistic possibilities.[1]

Tillich, as we saw within his theology of culture, differentiates three stylistic types. (1) Naturalism is form-dominated because it is based on the artist's direct encounter with objective reality. It presents elements of average experience—bodies, movement, line, color—in a recognizable manner. (2) Idealism is not bound to encountered reality. It brings out the potential perfection or beauty in persons and things which are not readily evident. The ideal potentialities are real but not yet actual. Form is not dishonestly "beautified," but the potential is realized in this balanced-type style. (3) Finally, the expressive, import-dominated style expresses reality's depths symbolically.

To sum up, naturalism focuses on surface reality, idealism on what surface reality might potentially be, and expressionism on depth-content. While all art is essentially religious in its expression, some is explicitly so through its expressive style.

Tillich lists characteristics of an expressionistic style which helps us to understand this stylistic type and his theology of art. These characteristics include space, movement, representation of the individual, nature, and finally, light and color.

1. In expressionism, space is not presented in perspectival form. It is disrupted, even dislocated. Tillich cites the archaic sculptures of the Greeks, since they are presented against the temple wall and are thus not meant to be walked around but approached only directly. Three-dimensionality is thereby denied. In expressionist painting, the artist does not attempt the illusion of three-dimensional space because this illusion is less expressive than disrupted space or a space not common to our experience.

In 1933, Tillich gave an address on "Dwelling, Space, and Time," at the dedication of a house in Potsdam, which helps us comprehend his perception of space. Space and time stand at the philosophical pole of ideas, while dwelling is on the concrete normative pole. While space is a comprehensive abstraction of our reality, it does not exist as a thing or even as a container for things. Dwelling is the concrete realization of space. It is small, concrete, and everyday; it is the human being's first and most immediate relation to space. Dwelling is the here-and-now where we discover and interpret meaning. In making dwelling, humanity creates space. In the following passage Tillich defines space.

> Space is the manner in which living things come into existence. Space is the power over space, the power of living things to create space for themselves. There is no space in itself; rather, there are as many kinds of space as there are ways in which living things create space for themselves, and that means, ways in which life becomes actuality. It is by spaciality that everything living, including the human, is recognizable.[2]

No space exists, *per se*, but everything, including the human being, may be described according to the way in which it is related to space. Several levels of living things exist, so Tillich describes four types of space.

1. The first type is the inorganic in which things exist alongside each other. Each fills its own space beside another's filling of space. The other is warded off, not included. The other is moved from without with pressure. Inorganic things are impenetrable and in opposition. No unity of space is found. The walls of houses and rocks are inorganic.

2. Vegetative or organic space dissolves the rigid opposition of the inorganic as it thrusts itself forward beyond itself. This type of space, common to plants, unfolds and evolves as it maintains its center and self-unity. The life process of growth illustrates an interpenetration of things standing alongside each other.

3. Mobility is the sphere of the animal. The inorganic occupying of space and the organic unfolding of space remain while adding movement. The animal is able to leave the soil, anticipate arrival at a distant space, and thrust itself into a non-immediate space. On the opposite pole, the animal wants limited space, its lair. Having been thrust out of the

supporting womb into space through birth, the animal longs to return to an enclosed and supportive space.

4. Finally, the human sphere includes a creation of all the above spaces as well as a twofold direction outward and inward. Outwardly, the human being creates the polarity of finite and infinite space. The finite space is the limited space in which the human being at times remains and from which the human being pushes forward into the infinite. The infinite is the human being creating space which breaks through every boundary of limited space. This is the human expression of power and existence.

Internally, because the human being is never content with what is, s/he pushes beyond to a vision of what might or should be. The human being consciously creates a world beyond the existing world; s/he shapes the space that is supposed to belong to him/her. This internal polarity is evident in Tillich's description of the theoretical and practical spheres of culture. Recall that in the theoretical pole the human endeavors to take the object into the person, to conform the person to reality, while in the practical pole the human being endeavors to inject itself into reality, to transform reality into what the person judges it should be. This internal pole is also evident in the stylistic types of naturalism (is content with reality) and idealism (envisions what should be).

The human being exists within a polarity that is decisive for the human shaping of space. On one pole, like the animal, the human being yearns for the enclosed, supportive space in which to put copies of vegetable and animal space by way of decor. At the other end of the pole, the person desires a detachment from limited space, to relate with infinite space, to take nothing in except what is necessarily connected with the meaning of the dwelling. We open ourselves wide to cosmic space and all forms of outlying space, and like light and air, they stream into our limited space. Whereas the animal impulse is to return to the lair, the human impulse is to move out into the infinite.

Because human existence is described in its relation to space, space has an essential holiness. At one pole, space is supportive. It is what binds us to the soil and gives us existence. House gods are the gods of the earth on which the house stands, and, often enough, blood is shed and sacrifices are made for them. Not only does a person's house limit space, but so does the neighbor's house, and the city, the landscape, the nation and its soil; these participate in the sacredness of space that supports our existence. At the other pole, the human space-creating power that pushes beyond limited space and seeks a unified dwelling place for humanity detaches itself from every special territory. Human-space creation becomes the creation of ever-growing living spaces; it becomes the creation of space in the progress of time. Time gains power over space.

Inorganic space has no relation to time other than that it endures in time. Organic space has a more intimate relation to time through its life process of youth, maturity, and old age. Animal space remembers the past, anticipates the future, and exists in the present, creating the threefold character of time. In human space, time becomes infinite as does space. The human being can anticipate a million years hence as easily as a million years ago, recognizing the possibility of each without end or beginning. Humanity thrusts itself beyond all form toward novelty and arrives at the boundary between time and space.

Time cannot annul space; it gains presence in space. In the present, and only in the present, are space and time united. To have space is to have presence; a lack of space drives one toward the future in order to create a presence out of it. Even those who have space are driven toward the future, for the future is the mode that makes time become time. To create time means to create a future which in turn tears us out of the present, out of space. Human beings are thrust out of space for the sake of time. When we are born, we are pushed from our limited space so that our time of existence may proceed. For Tillich, our deepest cause of struggle is that our house gods, the powers of limited space, resist being torn out into a more encompassing space, a space for humanity and a future into which human beings may be fulfilled anew. Time and space struggle against their unity in the present.

In creating space, humanity is called to balance the tension between the desire to dwell in a limited protective space, and the ache to thrust beyond the supportive and constrictive space into infinite space.

Susanne K. Langer makes space the pivotal concern in her understanding of the plastic arts. In her aesthetics, the aim of the visual arts is to articulate form in a wholly visual way. In so doing, the plastic arts present a shape in space that *shape* space.[3]

For Langer, like Tillich, space is not a concrete entity but a relational experience of our everyday lives. Space, again, is an abstract idea, realized concretely through our selves—our bodies moving and doing. Only through bodily experience can we discern spacial relations. Yet space eludes our attempts to shape and organize it because it is a dynamic and infinite field without concrete form.

Tillich is correct in pointing to the denial of three-dimensionality of space within the artistic expression of expressionism. But this denial is not specific to one style of art. All art denies three-dimensionality in that art does not shape experiential space but "virtual space." The artist not only organizes pictorial space through line, mass, and color, but also creates space. The artist creates a virtual space that exists for vision alone. The space created is independent of ordinary experiential space, which is known by touch, motion, and hearing, as well as vision.

According to Langer, virtual space is self-contained and complete in its visual structure. As experiential space is known through our relation to it, so virtual space is known through the individual forms in visible relation to one another within the created artifact. Spacial relations that we usually know through touch, hearing, or movement are presented through visual substitutes. Being only visual, the created space has no continuity with experiential space. It is cut off by its frame, and is void of any connection to realized space, even the side-by-side quality of inorganic space. The space created is significant in itself and it completely engages our vision in order to express meaning. It does not rely on its surroundings or on us for its achievement as art. The created virtual space is entirely self-contained and independent.

In many ways the artist makes space visible. This is evident in the multiplicity of artistic form. Different modes of virtual space lead to the distinct realms of painting, sculpture, and architecture. The pictorial artist as painter transforms a two-dimensional canvas or paper into a virtual space that presents a scene to our vision. A scene is not for Langer the representation of scenery but "a space opposite the eye and related directly and essentially to the eye."[4] This may be a complexity of forms, as found in the works of Hieronymus Bosch, or an ornamentation of forms common to Hindu art, or a color field found in Abstract Expressionism. All create scenes through the shaping of virtual space.

Sculpture, on the other hand, creates not scene but volume. The idea of virtual space is not only appropriate to the pictorial arts but to sculpture as well, even though it actually is three-dimensional in form. Tillich turned to sculpture to describe the denial of three-dimensional space! For Langer, space is amorphous. It can only be known to us by abstract mathematical formulas and through our relations to other data in our environment. It cannot be shaped or defined as a concrete entity, but art does shape space as it creates virtual space. Art makes space visible, including sculptural form. Sculpture commands the space it actually occupies as well as the empty space around it. The fundamental principle of sculptural volume is that sculpture makes the surrounding empty, invisible, shapeless space visible though the creation of an actual shape.

Sculpture is a semblance of organism reminding us that a living entity inevitably strives to achieve form. An acorn becomes an oak or a caterpillar, a butterfly. All of the parts of the form achieved are necessary to the functioning whole; if the parts cease to function the whole no longer lives. While sculpture is not a living organism, "its form is the form of life, and the space it makes visible is vitalized as it would be by organic activity at its center. It is virtual kinetic volume, created by—and with—the semblance of living form."[5] This is not to say that sculpture must be representational. Art, according to Langer, expresses the inner life or feeling, and thus it is not biological life that is expressed but "biological feeling." Since sculpture deals with volume and not scene, it is given over to touch, and so it makes "tactual space visible." Yet handling the sculpture is never enough; we have to step back to visually engage in the sculpture and its environment, a space that is independent of our space, a space that is all its own.

A dwelling, a place of limitation and support, is also art, the art of architecture. Utilitarianism and aesthetics are often confounded and even seem in conflict. Architecture is commonly spoken of as the art that creates and arranges space, but its unique quality is the articulation of an "ethnic domain" or virtual place made by treating an actual place. This domain is not a thing among things but a functional realm made visible. While cultural functions are a culture's ingredients, architecture, as a cultural creation, makes its image visible. The architect creates a culture's image, "a physically present human environment that expresses the characteristic rhythmic functional patterns which constitute a culture."[6] Architecture made in actual space is not continuous with that space, but is its own limited space. Its illusion is not tied to its place, its location. It may be moved or destroyed, but the place remains. Houses are rarely recognized as great art; they serve individual domestic needs. Great architectural ideas arise through the collective and public domains such as the church, the fortress, or the theater. The public realm is made visible and often dominates the community. These are not for habitation but for gathering to commune with God and humanity.

Tillich has approached space, in the above cited lecture, from the perspective of a culture-theologian. Space gives meaning to our existence. Our creation of space is meaningful as it gives character to our existence. Through our creation of space meaning is disclosed and our existence finds interpretation. In architecture, the window, garden, balcony, or other opening is pivotal to interpret this significant cultural creation theologically. The building limits and defines space, but the window opens to infinite space. That element connects limited space to the infinite. A contemporary building whose walls are primarily windows reveals a fear of limitation, a thrusting toward the infinite pole. A medieval castle, with its narrow space and high narrow windows reveals a fear of the infinite; it is a movement toward the pole of limitation and protection.

That Tillich and Langer are allied in their understanding of architectural space is evident. Both recognize architectural space is a cultural creation that expresses the character of that culture's existence. It makes that character visible so that meaning may be disclosed. Through architecture we are invited to be at home in our existence because it creates a semblance of our world.

In Langer's thought, art creates an "illusion" of virtual space: in the pictorial arts this illusion is "scene," in sculpture, "kinetic volume," and in architecture, the "ethnic domain." The space created is entirely independent and self-contained while it engages our vision in order to express its meaning. Langer uses an idea of organic space reminiscent of Tillich's to describe sculpture's virtual kinetic volume. Sculpture's space unfolds beyond itself as it maintains its center and self-unity. The sculptural space created is virtual because it is not organic in actuality, only within its intrinsic sensuality. Sculpture creates a semblance of Self, the center of life unfolding beyond itself.

What about the pictorial arts? They cannot be inorganic because they do have an inner unity that is inviting, rather than impenetrable and in opposition. Like architecture and sculpture, we are invited into the space that the pictorial arts create. But the scene created is not the kinetic volume common to organic space. Our clue here, and for all the arts, may lie in a fundamental difference between Tillich and Langer. Tillich states that space does not exist in itself but is created by all living beings, in all their variations, so that everything living may be recognized. Langer maintains that since space does not exist in itself, it cannot be created and arranged except virtually so through art. Art expresses meaning in the creation of virtual space. But Tillich the cultural-theologian states that all discernable space is created and that meaning for all living beings is disclosed in the process of creating space.

Architecture is the creation of a limited yet sustaining space. It is not totally enclosed but in some way open to infinity. The way architecture limits and opens to the infinite discloses its meaning. Sculpture moves beyond itself as it maintains its center. This, too, is a creation of space that evinces the limited-infinite poles. The sculpture is bound by its form; form is both sustaining and limiting. Yet sculpture thrusts outward from this form into the infinite. Sculptural space is not contained, but pulsates outward from its center. It is framed by a "window" that relates the limited and infinite poles. The archaic sculpture that Tillich experienced as defying three-dimensional space is true to all sculpture in that sculpture creates a virtual rather than an experiential space. Sculpture creates a semblance of organic life without itself being organic.

When sculpture is wedded to architecture, like the piece cited by Tillich, the sculptural form is at home in its dwelling space. Its limitation is stressed as it rests in its home. Simultaneously, its movement toward the infinite is thrust forward away from the architectural home. Unlike a free-standing sculpture, whose kinetic volume is evident in-the-round, sculpture wedded to architecture is at home while its kinesis thrusts beyond its limited architectural space. This suggests a denial of three-dimensionality. Architectural sculpture is like a living being standing in the doorway: on the border between the limited and the infinite. It has the potential of being one of the most powerful art forms because it makes visible the door, the border situation between the ambiguous poles of limitation and infinity. Yet while all architectural sculpture discloses the door situation, different styles of art will lean toward one pole or the other. The ecstatic sculpture of Bernini seems to flee its architectural space, while the archaic sculptures of the Greek temples emphasize a "being-at-homeness" in their architectural space. I believe this power of making visible the door, the border situation between the self-world relation, leads Tillich to describe space as a characteristic of expressionism.

We have yet to discover the role of the pictorial arts in Tillich's understanding of space. Since the pictorial arts create a virtual space of scene, they do so through the same elements of form, content (Tillich's definition), and style common to all the visual arts. While architecture makes our ethnic domain visible because it creates an experiential space of limitation and infinity, the pictorial arts invite us into a virtual scene that is bound by its aesthetic form but free in its content. The spacial polarity of limitation and infinity are operative in the pictorial arts as well. Art is a human creation and thus will reflect the human experience of space. For Tillich that is the tension between the desire to limit our space as a protection from infinity, and the desire to thrust beyond the supportive and constrictive space into infinite space. Indeed, art makes this tension visible through the creation of virtual space so that we may grasp its meaning.

An example will clarify this spacial concern. The Christo *Umbrellas* (Plate 7) installed simultaneously in Japan and the USA in October 1990, dramatically demonstrates the issue of space in the visual arts. In the USA, Christo positioned his *Umbrellas* on Interstate 5 just north of Los Angeles on the Tejon Pass. By locating 1,750 yellow umbrellas along an 18-mile stretch of California freeway, Christo forced us to experience his art. The work was not on a side road for us to choose to meander down in order to experience art, but was on a major highway that connects northern and southern California. He broke through the aesthetic callousness of a people hurrying-here, hurrying-there as they were thrust into an altered space of freeway.

The space that we have created called "freeway" has defined our existence in terms of fast and isolated movement. We do not notice the space as we rush through it. Then for a window of time, three weeks, an artist creates a space within an experiential space so that we may recognize space and know its meaning. When people entered the space of Christo's art, time so tied to space was also altered. The cars slowed down, the fast road became slow. The common experience of time as future, as *kronos* or linear time, ceased, and we entered a virtual time of *kairos*, real time, where time and space are united in the present.

In his creation of virtual space, Christo created a secondary quality of virtual time. People stopped, got out of their cars, and talked to one another. Isolation gave way to community. We were thrust from our limited space of cars into the virtual space of art where the tension of space, the tension of the polarities between limited and infinite space were visible. Simultaneously, the viewer was pulled to the dwelling, to stand under an umbrella in protection, and to regard the vastness of the California landscape which so effortlessly housed, even dwarfed, the 1,750 twenty-foot yellow umbrellas. Ironically and tragically, an individual did not find protection under an umbrella but death instead. On 26 October 1991, a Camarillo woman was killed by a wind-blown umbrella. This accident heightened the unpredictability of nature that Christo was presenting to our vision.

Christo employed the landscape as if it were canvas and painted umbrellas throughout in line and randomness, solitary and grouped, open and hidden, up close and far away, to create a virtual experience of space that was scene, kinetic volume, and ethnic domain. He made space visible so that we could enter into it and recognize it as true, so that we could find meaning in our existence.

2. Tillich states that immovability is the second characteristic of expressive style. As in the case of the archaic sculptures, Tillich believes that the artist is not interested in the relationship the figures have with one another but with the power of being within each figure. The artist is "not concerned with the realm of objectivity, the categories of time and space. He [*sic*] is interested in the expressive power embodied in the power of being which such a culture (in this case Greek) represents."[7]

Again we encounter the issue of time and space. In Langer's description of the plastic arts the creation of a virtual space (and thus time) is true for all the visual arts. The virtual space created is not concerned with experiential space but is its own type of visual space. It is independent and self-contained. This creation of space Tillich describes as not being concerned with objectivity because it is not an actual space relational to all other beings in space. Art is not dependent on its surroundings to gain its meaning, and thereby, all art is expressive, and thus for Tillich, religious.

3. The way the individual is represented is the third characteristic of an expressionistic style. Naturalistic art concentrates on the individual and its unique character. Rembrandt's work is exemplary of portraying the individual as interesting in himself/herself. Expressionistic art typologizes the individual. In it the artist prefers to express a type instead of depicting those who are merely interesting as individuals.

In 1920, Tillich wrote an essay entitled "Mass and Personality"[8] in which he analyzed types of mass movements within culture. He approached his subject by addressing the treatment of the mass found in paintings of several historic periods. Though Tillich is concerned with the treatment of the individual as an expressive characteristic, his analysis of the mass is an example of how the individual is represented, especially since he considers the relation of the mass to personalities of leadership.

The first period Tillich examines is the early Gothic of the high Middle Ages (1200-1300). Everything individual and of the crowd is dominated by an overarching idea: discipleship (as in a "Carrying of the Cross"), astonished adoration (as in a "Nativity"), or battle (as in a "Slaughter"). The individual is a type, each member of the mass being of that type. The presentation of space is two-dimensional, not three-dimensional which would accentuate the individual. The leader is the center of gravity without personal attributes. His formal position in the painting, large size, and rich adornment, highlight the leader as the mediator of revelation who gives the depth-content to the whole painting. This is the "mystical mass" in which the hierarchy of ideas, beginning with the deity, is what is real in the world and what arranges the leader and the led. Neither the mass nor the personality are separate realities. This suppression of the individual in favor of the overarching, supranatural (Tillich) idea describes Tillich's theonomous culture.[9]

Next Tillich considers the late Gothic or early Renaissance (1300-1450) of Germany and Holland where everything is changed. The individual is discovered and portrayed naturalistically within a presentation of three-dimensional space. Limited communal space has given way to an infinite space that accentuates the singularity of individuals, be they grouped in a scene of "Mockery" or a "Peasant Festival." Even the leader is one among many. The cohesion of the painting is a psychological homogeneity of the mass with a common emotion such as hate, pain, or desire expressed. Early Northern Renaissance art is a nominalistic period in which the individual ascends the throne and social revolution begins, notably the Peasants' Revolt. The "realistic mass," an autonomous culture, has replaced the theonomous culture of the "mystical mass."

In Baroque art (1600-1750), an "Erection of the Cross" or a "Calvary Battle" by Rubens (1577-1640), gives the mass a new significance when a surface linking of individual passion gives way to private personal convictions of faith. Humanity and nature are joined as the leader rides, participates in, and is supported by the crest of a dynamic wave of agitated depths churned by the collapse of the medieval, universalistic unity. The mystical monk supports the Gothic period, the peasant is central to the realistic period of late Gothic and early Renaissance, and the urban and aristocratic society sustains the Baroque. Here is a "dynamic mass with an aristocratic culmination that fulfills the most agitated time of European history."[10] This import-dominated romanticism is a transitional period between autonomy and theonomy that gives culture its dynamic quality.

Tillich analyzes modern art through the two styles of Impressionism and Expressionism. In Impressionism, the Baroque inner vitality of life gives way to a light which unites the mass externally. In works such as Monet's *Gare Saint-Lazare* (1877) or Degas's *Absinthe* (1876) the individual is subordinated to the surface appearance of nature. The substantiality of the mass is dissolved into a study of light, color, or movement; every object, every being is a vision of the moment, interesting and piquant, a piece of landscape. Form is everything in this form-dominated style. All reality is a mere object of formal technique. This is a time of social reform and Marxism. This mass is a "technical mass," a mass as object of impression and technique. That mass and nature are subordinated to artistic technology is evidence for Tillich of a thoroughly autonomous culture. Form takes primacy over any type of expression of metaphysics or depth-content.

The first decade of the twentieth century ushers in the new revolt of Expressionism. Outward form is still dissolved, but rather than for superficial technical aims, this new style aims at a deeper, more substantial meaning of life. Form is broken-through in order to unearth metaphysical significance. Like the Gothic, the mass is again of a type and the presentation of space is two-dimensional. Yet the spirit-bearing human face is prominent. The leader is missing and can only be born from the depth of the mass's longing and suffering. Tillich explains:

> A torpid something weighs on these masses, a metaphysical lack of redemption and longing for redemption. For that reason they are sometimes approximated to animals, then again elevated beyond humanity to ecstatically visionary heights.[11]

In Impressionism the object was everything, in Expressionism the mass is subject. Expressionism is the period of world war, the Russian "soul-chaos," and communist martyrs. A new mystical mass is revealed in Expressionism, the mass of immanent mysticism. Whereas the Gothic mystical mass was directed from above by a supranatural hierarchy, this new mystical mass breaks out of the depths of the soul that is vitally linked to reality. Expressionism announces and reveals the dawn of a new theonomous age.

In the analysis of Expressionism, Tillich does not cite examples of paintings as he does in the previously discussed periods. His description is general, even vague. What is the "torpid something" that he mentions is weighing on the masses? Only in his discussion of Expressionism does he mention Christian symbols. They re-emerge but are merely symbols that lack the power of those symbols that arise out of the depths of the masses.

Tillich has, however, discerned three types of mass. Mass is for him a linking of people who give up their individuality in favor of a quality that unifies them as a mass. This quality may be the supranatural of the

Gothic; psychological of the late Gothic and early Renaissance; inner piety of the Baroque with an aristocratic spirit; technology of the Impressionist; inner seeking and suffering of the Expressionist; or another variation. But in each case the mass falls into one of three types: (1) The form-dominated technical type in which the mystical principle is absent so that external powers, like natural instinct or technique, create the character of the masses (subjective-Impressionism, objective-realism—late Gothic, early Renaissance). (2) The subjective import-dominated dynamic type in which an inner principle struggles with motive power for configuration (Baroque). (3) The objective import-dominated mystical type in which the mass is pervaded by an immediately spiritual, mystical principle (Gothic, Expressionism).

When the three types of masses are connected with Tillich's typologies for religious art within a theology of culture, Tillich lumps together the poles of the form-dominated type to realize the technical mass, and separates the poles of the import-dominated type to describe the dynamic and the mystical masses. The balance type (idealism and classicism) is omitted altogether. In idealism and classicism the figures emphasize individual characteristics. This type is problematic for Tillich's scheme of the artistic treatment of the individual or the mass. Tillich prefers a typology that expresses something other than individual characteristics. Styles that create types based on individuality do not fit his scheme. Rather than struggle with this problem, Tillich fails to mention this type of art. However, he does cite classicism as an example of an expressive style's treatment of two-dimensional space.

In 1946, Tillich wrote an essay, "The World Situation," in which he includes a deliberation on portraiture. He begins his discussion with Rembrandt (1606-1660). Rembrandt's portraits are self-enclosed worlds that bear the marks of unique personalities and histories. They express the ideals of a humanistic Protestantism. Personal existence is determined by life experience with its tragedy and ultimate hope. Community is formed as active groups unite for a common purpose. This community is future-oriented because its bonds are based on common devotion to particular aims for which it is necessary to fight. The common ground of universal authority has dissolved.

Giotto's (1266-1337) portraits of St. Francis (Plate 8) and other monks, which are at the other pole, express the Divine power that possesses humanity. This religious expression overpowers individual character and personal experiences. St. Francis is more a type than an unique personality. Personal life found its meaning and center through its relation with transcendent reality. The community overarches individuality as all participate together in a loyalty to transcendent reality.

In the portraits of Titian (1487-1576) the individual expressions of humanity (greatness, beauty, power) are represented. Meaning is furnished through belief in the divinity of humanity and the humanity of the divine. Outstanding individuals are predominant. They represent general humanity, but are isolated individuals rather than communal participants. The transcendent quality that overpowers Giotto's figures has vanished, yet Titian's portraits are not the individual personalities of Rembrandt either. Rembrandt's portraits, still subject to absolute forces and shaped by the Protestant conscience, stand independent of transcendent grace and humanity. "In these three painters, the development of the ideal of personality in the modern world finds classic expression."[12]

By the middle of the nineteenth century, individuals with highly developed intellect and strong will appear. Personality rules the form. "Willpower and technical rationality are united, and thus the way is prepared for the Fascist type in which the last remnants of the classical and humanistic ideal of personality are completely lost."[13] The naturalistic personality, formed by modern technology and economy, is devoid of divinity and humanity, and ultimate meaning is gone. Both the common spiritual ground of Giotto and the common spiritual purpose of Rembrandt are lost. Community has disintegrated into masses with neither a common ground nor a common purpose.

While in 1922, Tillich defines a mass as a linking of people who, in joining, give up their individual qualities in favor of a quality of the whole, by 1946, the mass has become the consequence of individualism and the resultant crumbling of community.

Tillich's discussion of portraiture is interesting because he cites three individual artists instead of general historical periods, although at the end of his lecture he makes sweeping generalizations about the naturalistic personality of the mid-nineteenth century with nary an example. In the essay on the masses, late Gothic and early Renaissance art present a form-dominated technical type of mass in which import is suppressed by a concern for natural form. The art is nominalistic in its discovery of the individual and three-dimensional space. The cohesion of the mass is due to a psychological homogeneity.

When Tillich looks at a specific artist, Giotto, we discover a different analysis. The naturalism of Giotto is overpowered by an expression of transcendent reality. The cohesion of the community is a common spiritual ground. Giotto's art is import-dominated. Granted, the "realistic mass" of the late Gothic and early Renaissance is from the north (Germany and Holland), and Giotto is from the south (Italy). It is possible—even probable—that geography could account for this difference between the northern and southern styles. However, Tillich, is quite consistent in his evaluation of art when he contends that the discovery of three-dimensional space and "the individual" results in a subordination of import to form, and a loss of overt expression of transcendent reality.

Giotto is known in art history for his extraordinary achievements in painting. He ushers in a powerful concern for naturalism and the Renaissance. Because of this fact, I am not convinced that this discrepancy in Tillich's thought is due to a geographic consideration of styles his tendency to generalize about art. When Tillich looks at a specific artwork, something different than what his generalizations suggest arises. The works of Giotto provide a concrete example of form-dominated naturalism, yet they do not conceal import (by Tillich's own admission) but the import transcends the natural form. Natural form can be the vehicle for meaning, ultimate concern, without being shattered.

In sum, for Tillich styles such as Gothic and Expressionism treat the individual in an expressive manner. The form of the individual is broken through by import so that depth-content is revealed. It is not interesting in itself, but is a window for the metaphysical. This may arise from within the individual, as in Expressionism, or be imposed on the individual as in Gothic.

4. After looking at types of individuals depicted, Tillich considers their relationship to nature. He is a little muddled in his description of this fourth characteristic of expressionistic style. Tillich claims that in the centuries when expressionism prevailed, humanity and nature participated together in a common symbolization. Then he cites a primary problem for the history of art: to show how a different relation to nature developed in late Gothic and early Renaissance art. The more naturalistic the art, the more humanity is a part of nature, standing within nature. Nature, then, is the embraced reality. When this is the case in art, Tillich is troubled. But how humanity and nature are depicted as participating together on the expressive pole, or how humanity is depicted as a part of nature on the naturalistic pole, we are never told. Tillich offers no examples or specific formal elements to help our understanding of this relationship in expressionism.

Tillich's understanding of the nature of the individual in portraiture and within the mass helps to decipher this fourth characteristic. Since the relationship of humanity to nature does not emerge as an art-historical issue until the advent of early Renaissance art, especially in the work of Giotto, I will skip over the Gothic period. Suffice to say that within the Gothic period a natural approach to the individual and nature is overpowered by the overarching idea. For Tillich, both the human and natural forms are typological or symbolic forms that impart import.

Giotto (1266-1337) and the early Renaissance (1300-1450) usher in a naturalistic concern for gesture, nature, and three-dimensional space. No longer are humanity and nature spiritualized. Both are interesting in themselves. Spirituality, however, is realized through the sensual experience. The supranatural no longer suppresses the natural; instead, the natural shares in manifesting meaning. Tillich, though, describes this transition toward the natural as form-dominated, nominalistic, and autonomous. He believes that naturalism is limited and assumes its own absoluteness, that depth-content does not break through art's treatment of the everyday, and that subject matter overpowers both form and content. So for Tillich, a religious perception of reality is excluded.

Tillich does not appreciate naturalism's ability to deepen understandings of our world. His metaphysics does not include explicit reality but shatters it. When human and natural form are presented naturalistically, import is suppressed in favor of form and the religious element is overtly absent. According to Tillich, only when natural forms are shattered is art expressive or explicitly religious.

When Tillich looks at Giotto's paintings, his experience of a naturalistic art form is quite different. Import does transcend naturalism without shattering its form. Three-dimensional space is an avenue for the expression of ultimate concern. Tillich defines Giotto's figures as typological and symbolic of transcendent reality like that found in Gothic art. Nevertheless, in viewing the works of Giotto, we discover the emergence of the individual in a scene of average experience. Giotto is a fine naturalistic painter who does not fit Tillich's categories.

The idea that naturalism is opposed to expressionism is re-stated in Tillich's discussion of Impressionism and Expressionism in "Mass and Personality." Impressionism subordinates the individual to the surface appearance of nature. The individual is a part of nature; nature is decisive. Impressionism is an autonomous art that excludes the expression of depth-content or import. Expressionism shatters form so that depth-content or import may be revealed. The individual and nature are not easily distinguishable, for their natural form gives way to an expressionistic one. The individual and nature participate together in a deeper symbology of meaning.

5. Finally, the fifth characteristic of an expressionistic style is light and color. Since Tillich rarely comments on light and color, the following passage is unique and worth including in its entirety.

The gold ground which is the symbol of transcendent light or the broken

> light of the stained-glass windows which gives us the manifoldness of nature, not as it is outside, but through their so-to-speak heavenly colors, represents an expressive style. But another step was taken by the preaching orders in the later Middle Ages and then by Protestantism, namely, the accent on daylight, the rational light, the natural light. Here the difference between an expressive and a naturalistic style is evident.[14]

Gold-ground and stained glass windows with heavenly colors are expressionistic. Daylight is naturalistic. Tillich has offered no clue as to what a heavenly color is, other than gold, nor has he even mentioned naturalistic color.

Expressionistic light is the actual light that comes through the window. Other than the mention of naturalistic light as rational and daylight, we are left in the dark regarding naturalistic color. Impressionism, with its emphasis on color and light is not mentioned, nor is Expressionism.

Light and color are primary, formal elements in all the visual arts. Art criticism and art history consider the use of light and color within painting, sculpture, and architecture as fundamental elements in their analyses of the arts. Yet, it is here that Tillich is the most vague. He does not consider issues of light and color such as intensity, direction, location, relationships, or value.

The only example of an expressionistic light is the window. In Tillich's description of space, the window is the pivotal element in a theological-cultural interpretation of architecture. The window connects the limited space of the building to the infinite space beyond. As the window opens to infinite space, it allows the infinite to stream into limited space. Thus, in regards architecture, stained glass windows are expressionistic.

To understand what constitutes an expressionistic style, we have looked at Tillich's characterization of space, immovability, representation of the individual, the relation between the individual and nature, and light and color. Tillich uses archaic sculpture to talk about space and immovability. His thought on architecture clarifies these features. He cites historical periods to talk about the representation of the individual and the individual in relation to nature. His essays, "Mass and Personality" and "The World Situation," illuminate aspects of the individual. He describes actual light to talk about light and color. Here he is too vague for us to critically evaluate.

Tillich classifies all of the plastic art forms but not holistically. One type of art is helpful in discerning one characteristic, another art another characteristic. Could these five characteristics be successfully and clearly applied to all types of visual art? A generalized statement about the visual arts must be consistent within all of the plastic art forms. Before attempting an answer to this question, we need to ascertain what for Tillich an expressionistic style is.

An expressionistic style is, according to Tillich, neither accepting of reality as experienced (naturalism) nor anticipatory of what potentially might be (idealism). Instead, it expresses the dimension of reality that cannot be expressed in any other way. It makes the artistic dimension of reality—import—visible. In spacial categories, naturalism makes the present visible, idealism makes the future visible, and expressionism makes the underlying struggle of time and space against their unity visible.

An expressionistic style employs natural form not as meaningful in itself, but as a vehicle of revelation through which import, both supranatural and implicit within, manifests itself. Both import and form are mutually immanent. An explicitly religious interpretation of reality requires the destruction of natural form under the power of the import expressed. Natural form is not completely obliterated, but its concreteness bears a deeper reality. Meaning is not innate in the objective structure of reality, as naturalism suggests. When the infinite penetrates our experience of a limited concrete reality through the creation of an expressive form, meaning is revealed. Thus, an artistic concern with three-dimensional or experiential space, individuality, or natural form is rejected by Tillich.

In religious expressionism, what is expressed and its means of expression are not parallel. This definition is so general that it can be applied to all art forms, since each presents its own reality which transcends a focal interest in the object. Tillich's definition is limited in its scope of what makes a style explicitly religious. Both naturalism and idealism, which affirm surface reality, are excluded as religious types of expression.

Tillich's definition of an expressive style leads us to consider the contemporary arts, especially the Abstract Expressionists of the 1950s in New York. Would the complete destruction of recognizable or natural form be the most religious style? Or would import dominate the form so powerfully that the tension between autonomous natural form and theonomous expressive form is lost? If this tension is dissolved then so is the religious element. Would the new form created in Abstract Expressionism be theonomous or heteronomous in Tillich's scheme? Since Abstract Expressionism is freed from the limits of concrete reality through a means of expression that is considered to have no relation to objective reality, is the vital tension between form and import dissolved?

In 1958, Tillich attempted to interpret the expressive style in the contemporary visual arts since 1900, with the advent of Cézanne. He lists seven characteristic elements in the contemporary expressive style.

1. The first characteristic is the attempt, begun by Cézanne, to produce paintings that stand by themselves without relying on natural form or subject matter to achieve artistic status. The painting is liberated from the "naturalistic bondage" to ordinary experience of reality, and from an "idealistic transcendence" that brings out an object's potential perfection. Form loses its relation to content.

2. The second characteristic is the preference of two-dimensional space. The three-dimensional reality of mere objects is removed. Two-dimensional space in sculpture is achieved through the art of reliefs, through wedding sculpture to architecture, and in contemporary art, through the use of empty space between the sculptural elements. The flattening of space reveals the expressive style's affinity to religion.

3. The disruption of the surface of our ordinarily encountered world

is the third characteristic. The concrete structure of reality is lost to the urge for pure expression. Picasso's *Guernica* (Plate 9) is exemplary of a disrupted surface. Bodies, facial features, animals, and technical structures are torn from their ordinary context and are put together in a new way. The painting presents us with a configuration of reality that we do not experience within ordinary reality.

4. The fourth characteristic is the psychological depiction of the brokenness of humanity's rational approach to the world as found in the works of the surrealists. Subconscious images are made visible and brought to consciousness. Objects of ordinary reality are reorganized in a fantastic way. The familiar becomes strange, and the general phenomenon of estrangement is made visible.[15]

5. Fifth, natural reality is dissolved to reveal its underlying structure—cubes, spheres, cones, lines, and planes. The Cubism of Picasso (1881-1973), Braque (1882-1963), and Gris (1887-1927) employs geometric elements to reveal the elementary power of being. These painters analyze the structure of being in order to construct meaning.

6. The sixth characteristic is the consequential disappearance of organic form, especially the human face. Only the structural elements of inorganic form are recognizable. These forms are endowed with the human power of being and meaning. This art presents a disrupted and objectified world and in so doing it conquers disruption and objectification.

7. The seventh characteristic is non-representational art. Elements of encountered reality give way to the relation of colors in an unordered-ordered geometric network. The physical materiality of painting becomes the painting itself. While reality's materiality and inner relations are not transcended, some expressive power is revealed.

Abstract Expressionism is not the most religious style; it has lost the explicitly religious element because its style is extreme. The expressive power it shows is the basic expressive power common to all art. Abstract Expressionism is religious in the larger sense that all art is religious, but not in the narrow sense of Expressionism. Tillich places Abstract

Expressionism on the heteronomous pole because it makes the materiality of art ultimate and despises natural form.

Tillich is consistent in his analysis of an expressive style. He has an affinity for two-dimensional space and abstracted natural form. Recall, all art abstracts form. The term "abstract" here refers to images of reality depicted in a non-representational way, that is, the internal or external relations, colors, mass, etc. are not common to our average experience of the realities portrayed. For Tillich religious art must be balanced on the pole between natural form on one side and non-representational form on the other. To err toward the end of either pole would be to dissolve the tension of the poles and thus dissolve the religious element.

Within the seven characteristics of a contemporary expressive style, only paintings of Cézanne and Picasso were cited. Braque, Gris, and the Surrealists were mentioned in general. The second, sixth, and seventh characteristics are lacking examples. Again, Tillich generalizes about art without supporting evidence, leaving the reader dissatisfied and often confused.

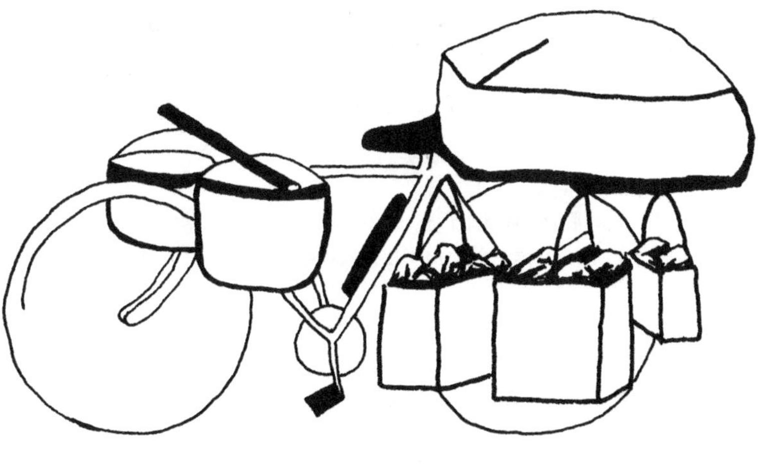

NOTES

1. Paul Tillich, "Visual Arts and the Revelatory Character of Style" (1958), in *OAA*, p. 130.

2. Paul Tillich, "Dwelling, Space, and Time" (1933) in *OAA*, p. 82.

3. Suzanne K. Langer, *Feeling and Form* (New York: Charles Scribner's Sons, 1953), p. 71.

4. Ibid., p. 86.

5. Ibid., p. 89.

6. Ibid., p. 96.

7. Paul Tillich, "Art and Society" in *OAA*, p. 35.

8. Tillich, "Mass and Personality" was later published in 1922 as *Masse und Geist: Studien zur Philosophie der Masse* in *Volk und Geist*, no. 1 (Berlin, Frankfurt: Verlag der Arbeitsgemeinschaft, 1922). My study is taken from "Excerpts from Mass and Personality," trans. Robert P. Sharlemann, in *OAA*, pp. 58-66.

9. I am indebted to Michael F. Palmer, *Paul Tillich's Philosophy of Art* (Berlin and New York: De Gruyter, 1984), pp. 105-106, for the connection of "Mass and Personality" with Tillich's ideas on autonomy, heteronomy, and theonomy.

10. Paul Tillich, "Mass and Personality" in *OAA*, p. 61.

11. Ibid., p. 63.

12. Paul Tillich, "The World Situation," in *The Christian Answer*, ed. Henry P. Van Dusen (London: Nisbet and Company, LTD, 1946), p. 10.

13. Ibid., p. 11.

14. Tillich, "Art and Society," p. 35.

15. Tillich, "Visual Arts and the Revelatory Character of Style," p. 137.

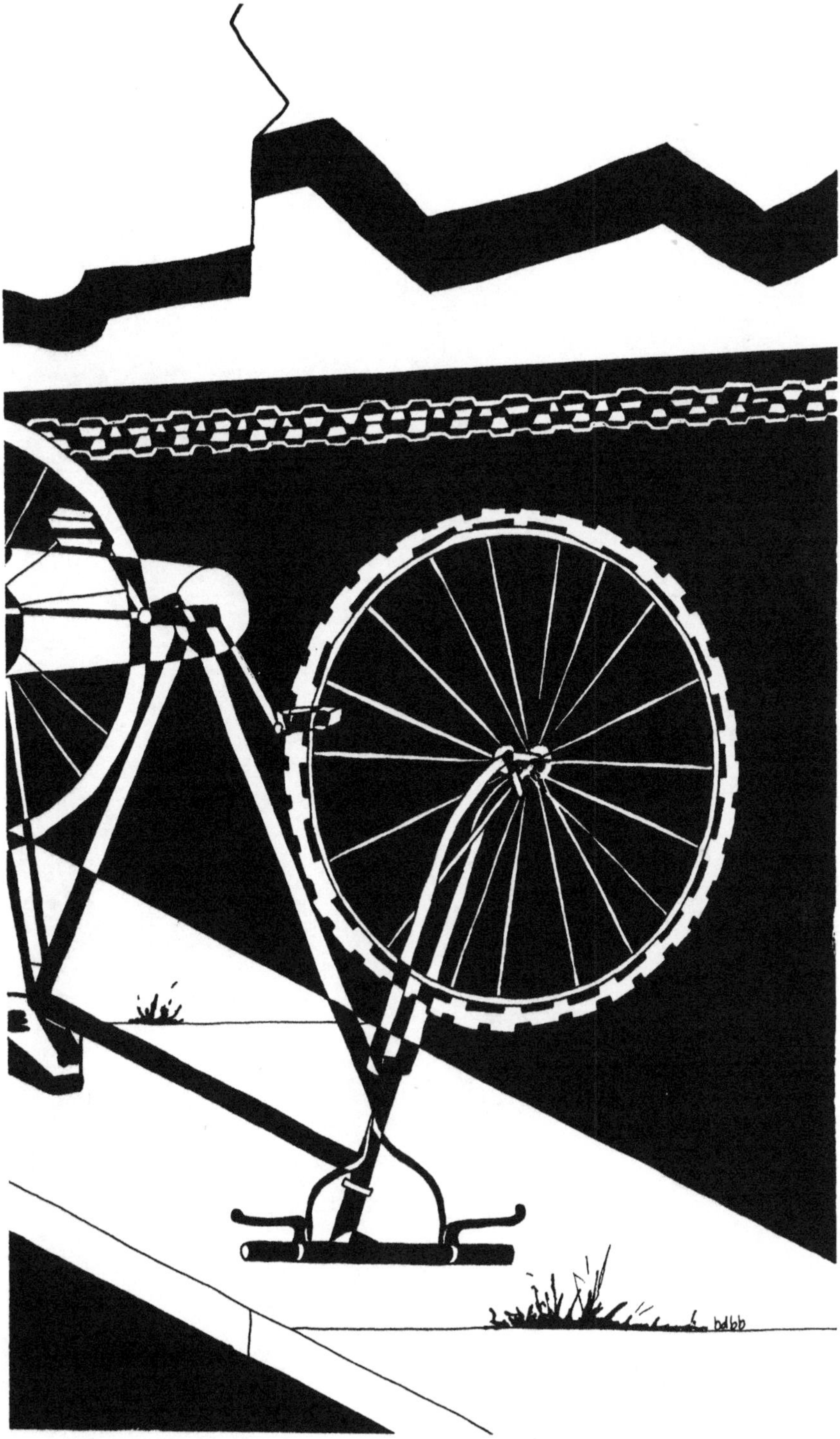

Chapter Nine

Style and Content

The third level of the relationship between art and religion is the realm of religious content or subject matter. Tillich here is concerned with the question of whether we can have religious art if the painting is religious in subject matter but not in style, that is, if it is not expressive but naturalistic. Recall that naturalism is concerned with the surface reality, while expressionism probes beneath the surface into the depths. Tillich sees a great incongruity between style and content when religious subject matter is painted naturalistically. A contradiction exists between a focus on natural reality and its underlying meaning.[1]

From his interest in the contemporary arts, Tillich continues the question, wondering if a painting that is non-religious in subject matter but of an expressive or religious style can be religious art. He moves beyond his adage that all art is essentially religious to a concern with art that is "existentially religious." His predominate question is, does a particular art "speak to its time?"

Tillich's concept of religious symbols is informative. A symbol, like a sign, points beyond itself. A stop sign points beyond itself to the stopping of the car; a religious symbol points beyond itself to the divine. Yet the symbol is not intentionally produced but grows out of the relation between what is symbolized and those who acknowledge the symbol as symbol. Unlike the sign, the symbol participates in the reality it symbolizes.

A religious symbol grows out of the divine-human relationship and participates in that relationship. When that relationship changes, or takes on new meaning, the symbol will die and a new one will emerge. The symbol is grounded in the power of what it symbolizes and is revelatory of truth. It reveals meaning, realms of existence that cannot be revealed in any other way, and it opens up the soul.

Symbols reveal to us something explicitly that we may only know tacitly; at the same time, symbols employ aspects of reality to provide form for what transcends everything finite. The American flag employs red, white, and blue cloth to give concrete reality to the concept of the United States. The finite realm of fabric is taken up into the conceptual realm of nationhood. The fabric itself takes on dimensions of "sacredness" as it participates in symbolization. A direct correlation between the flag and nationhood exists in the flag. The "flag" form gives it the general character of nationhood while its unique form of a red, white, and blue flag distinguishes it from other flag symbols.[2] Secularization is the separation of the finite symbol from its ultimate reality. When the flag is "just a cloth," its participation with the symbolized ceases and the interrelationship is cut.

Religious symbolism points to the transcendent and immanent realities of God, to being-itself. The immanent level of symbols are the sacramental realities that make the transcendent present in the world. God is not made a part of the empirical world; the empirical world is made a vehicle to reveal the ultimate ground of being that transcends all of life.

How the transcendent is made present is the artistic problem that Tillich describes in this third relation of art with religion. Art is a fundamental type of symbolization in Tillich's aesthetics. The form transforms content into a work of art. The artist uses the subject matter, natural form of ordinary experience, to create a painting that is not the natural form perceived but something new interrelated to both the form and its ground of being, or import. When the content of the painting is taken literally, it becomes "just a piece of cloth," and its power as a symbol is lost.

That a painter paints religiously, or paints religious subjects, does not guarantee religious import. To create religious substance, we must paint symbols that grow from our very being. They cannot be the symbols of another age, unless they continue to live in our depths and open up our souls. Expressionism is the style that expresses what is alive in our depths. Idealism and naturalism do not point beyond themselves; they both concentrate on explicit reality, not what transcends the surface. This is why they contradict religious content, and they do not make the transcendent existentially present. The transcendent can appear through any form, except these forms are derived from finite reality and thus can present themselves as finite where the transcendent is not identified. A naturalistic style tends toward the finite reality of form, whereas an expressive style overcomes the finite in order to bear the transcendent.

Recall, Tillich uses the term "content" simply to refer to an artifact's subject matter. In art criticism, "content" is understood to be an artwork's comprehensive meaning. "Iconography" is the term commonly used to refer to subject matter. Tillich, however, uses "content" to refer to iconography and "import" to refer to content. This is a cause for confusion among art-critical readers of Tillich's aesthetics.

From Polanyi we have learned that perception of subject matter and form requires a low degree of indwelling, whereas apprehending an artwork's content (meaning) demands openness and abounding participation in its event. During sustained segments of visual contact our sensual, emotional, intellectual, and spiritual world is drawn into and stimulated by the work of art. The mix of the artistic event, our selves, and our world reveals meaning or content.

While on a furlough in Berlin during World War I, Tillich had such an experience of content with Botticelli's *Madonna and Child with Singing Angels* (Plate 5). The ecstatic power of this encounter drove Tillich to respond creatively. He was carried away, and he responded as a viewer who sees in a new way. It was a truly creative moment for Tillich who can only reply poetically. Because of the importance of this event I will quote at length the account of his experience with the Botticelli painting. A poesis in his description carries the life of his encounter.

'One Moment of Beauty'

> Gazing up at it, I felt a state approaching ecstasy. In the beauty of the painting was Beauty itself. It shone through the colors of the paint as the light of day shines through the stained-glass windows of a medieval church. As I stood there, bathed in the beauty its painter had envisioned so long ago, something of the divine source of all things came through to me. I turned away shaken.

> That moment has affected my whole life, given me the keys for the interpretation of human existence, brought vital joy and spiritual truth. I compare it with what is usually called revelation in the language of religion. I know that no artistic experience can match the moments in which prophets were grasped in the power of the Divine Presence, but I believe there is an analogy between revelation and what I felt. In both cases, the experience goes beyond the way we encounter reality in our daily lives. It opens depths experienced in no other way.
>
> I know now that the picture is not the greatest. I have seen greater since then. But that moment of ecstasy has never been repeated.[3]

On a later occasion while touring the National Gallery in Washington, Tillich's eyes were caught by *The Dancing Couple* (1663) by the Dutch painter Jan Steen (Plate 10). Tillich, at the time, was looking for paintings with an explicit religious style, but he could not look away from Steen's painting.[4] Tillich was carried away by this "very dynamic, very vital" painting. He recognized the divine ground of being in this scene of peasant vitality.

Two paintings carry Tillich into a rich experience of content. Both are of the form-dominated realism type, and thus, according to Tillich, are extremely secular in character. How curious that these paintings, powerful aesthetic encounters for Tillich, are exemplary of painting that is merely implicitly religious. Indeed, these paintings probably led Tillich to state that all art is essentially religious. Their power and religiosity do not fit his category of expressionism—rather than struggle with this problem, Tillich seems satisfied to place them within a larger category that understands all art to be religious.

In the third level of the relation between art and religion, Tillich questions art that is incongruous in style and content (subject matter). He answers this question in a 1955 lecture. But in 1952, he sets this problem aside and offers instead an answer in terms of a fourth level in the relation between art and religion: subject matter and style are clearly expressive of ultimate reality.[5] Tillich suggests that the expressive style may transform the secular subject matter so that it may be deemed religious art—but this is merely a musing in 1952. He challenges the artist to answer his questions. This is for Tillich the realm of liturgical art where the art should be expressive in style and in content.

Tillich summarizes the four levels of the relationship between art and religion because they are critical to understanding the relation of religion and art.

1. First level, all art with its universally expressive power, even secular and semifrivolous art.
2. Second level, art which expresses the ground of being or ultimate concern, through stylistic characteristics that contradict reality as we encounter it in our daily world.
3. Third level, art with religious content, independent of any particular style, wherein the question arises of whether or not a naturalistic style is possible.
4. Fourth level, art in which religious subjects and religious style come together.[6]

In 1955, Tillich gave a lecture similar to the 1952 lecture cited above at a number of American institutions. In each lecture Tillich struggles with the preceding questions regarding the necessity of style and content (subject matter) to be in congruence with one another for religious art to be possible. He holds steadfast to the four levels of art as it relates to religion, but he contours them to meet his overwhelming concern with style and its relationship to content (subject matter).

Whereas in the 1952 lecture the first level claims all art to be essentially religious, and this remains true in Tillich's thought, in 1955 he streamlines this level toward "secular art." Ultimate concern is only indirectly expressed when both style and content are irreligious. The former second level is concerned strictly with defining an expressive style that directly expresses ultimate concern. The second level continues to describe the expressive style as religious, but Tillich answers his question on incongruity regarding style and content; the style is religious but the content is not. Since Tillich discusses the issue of non-religious subject matter in the new first and second levels, the third level continues to grapple with religious content but now definitively without religious style. The fourth level remains unchanged; the style and content are both religious. The following examples of each of these levels demonstrates the evolution of Tillich's categories.

1. Non-Religious Style, Non-Religious Content. This is exemplified by the art of Jan Steen, especially *The World Upside Down* (1665) in which the vitality of the self-affirmation of life is almost ecstatic. God is as present in this secular painting as God is implicitly present in a religious one. The power of being is indirectly visible when both the style and content are non-religious. This is the form-dominated type found in naturalism. Rubens's (1577-1640) *The Prodigal Son* (Plate 11) is also cited as portraying a cosmic sense of reality that draws the viewer in. But it is completely concerned with the surface reality of color and forms. These secular paintings reveal levels of reality that cannot be grasped in any other way. They are, thus, essentially religious, though they are secular in style and content.

Earlier we saw that for Tillich Rubens's art was exemplary of the transitional dynamic mass, and is categorized as an import-dominated romanticism. An inconsistency appears in Tillich's thought. Perhaps he changed his mind regarding Rubens over the thirty-three year period between 1922 and 1955; perhaps one painting of Rubens fits one category while another painting fits another category. That the work of one artist fits into more than one category affirms that Tillich was open to art's event and that he was not limited by his categories. However, I am troubled because the "Return of the Prodigal" is religious subject matter, and thereby does not fit this first level.

In my view, this level includes the type of art that many critics skip over, the academic period. Art in the seventeenth and eighteenth centuries, the work of Anton Raphael Mengs (1728-1779) for example, is no longer a part of the church. Knowledge, duty, and science take precedence over the vital life of the church. Artists no longer serve their art but command it. They search for themes rather than ideas. The effect is an art which is finite in its expression. The subject matter is not religious but its themes are of human grandeur, such as imperial portraiture. Humanism rules the form in which both the spiritual ground of divinity and humanity are lost.

2. Religious Style, Non-Religious Content, the Existentialist Level. Cézanne ushers in the existentialist movement in the visual arts, according to Tillich. All of life has become a still life; organic forms, thus idealism, disappear. Instead, the forms of existence are atomistic and disrupted. Modern artists embody the power of being in inorganic form, such as Cubism. The disruption of form common to modern art moves beyond the surface and its beautification. Organic unity is no longer an artistic concern. Modern art presents the underlying depths of being from which reality is constructed.[7]

Van Gogh's *Starry Night* (1889) and *Night Cafe* (1888), and the works of Munch (1863-1944) go below the surface to reveal the creative powers of nature and the horrors of existence. Picasso's *Guernica* (1937) (Plate 9) is, for Tillich, one of the greatest religious paintings ever painted. Picasso paints the horror of "saturation bombing" in a way that "the 'piece' character of our reality is perhaps more horribly visible than in any other of the modern pictures."[8] Picasso reveals the human situation without any cover: disruptedness, existential doubt, emptiness, and meaninglessness. The depths of estrangement and despair are revealed in a profoundly religious style, though the content is non-religious.

Other artists, such as Braque (1882-1963), Chagall (1887-1985), de Chirico (1888-1974), dissolve organic reality in order to move away from the surface to reveal God and the basic structure out of which God created the world. Each creates a work that is out of context with experiential reality so that the depth-content may be expressed.

The paintings in this category profoundly face the human predicament and the religious question that comes from it. Time, space, causality, and substance lose their ultimate power to matters of more radical import. These painters, and others such as Goya (1746-1828), Brueghel the Elder (1527-1569), and Bosch (1450-1560), give meaning to our lives as they express the displacement of humanity in the world.

They describe the human situation in its depths. Therefore, though the subject matter is non-religious, these expressionistic paintings are extremely religious in style. The artists disrupt the surface and make ultimate reality visible.

3. Non-Religious Style, Religious Content. Raphael's (1483-1520) paintings of Madonna and Child (Plate 12) are not religious in import or style. Instead of a disrupted reality, Raphael presents a reality of harmony and beauty. Harmony, according to Tillich, is non-religious in style. Artists such as Fouquet (1425-1480) and Rubens present beautiful or court women as Madonnas. How can such elegant women be symbols of the mother of God? The content may be religious, but these paintings of harmony and beauty are not religious in style. They are dangerously irreligious.

Tillich places much of Renaissance art which stylistically emphasizes human grandeur in this category. Yet, primarily for Tillich, this is the arena of kitsch, the unholy marriage between idealism and realism. It is dishonest in form, sentimental, and vapid, for example, Roman Catholic holy cards. Regrettably, most Roman Catholic sacred art since 1750 is just as bad. In *Disputed Questions*, Thomas Merton claims that this kitsch is damaging to our religious life. Art is powerful, and in its dishonesty it has a negative impact on our faith. Better no art, like the Protestants, than this![9]

4. Religious Style, Religious Content. The expressionistic form that disrupts the surface in order to express depth-content is operative in this level. Tillich only cites crucifixion scenes that fulfill this category. The Crucifixions of El Greco (1541-1614), Grünewald (1475-1528), and Sutherland (1903-1980) all are religious in style and content. Indeed, Grünewald's *Crucifixion* of the Isenheim altar (Plate 13) is, for Tillich, the greatest German picture ever painted.

The paintings of other artists, such as Nolde (1867-1956) (Plate 14) and Rouault (1871-1958), are more religious when the content is non-religious, than when it is purposely religious. This leads Tillich to question the value of traditional symbols as appropriate to the modern experience of existentialism. Perhaps these symbols are no longer true to the modern experience and no longer live in our depths. Authentic religious art in both style and content can only be a result of an inner necessity. He concludes that the Christian symbols are no longer a part of the vital life of contemporary culture and are thus unable to be expressed adequately in art. What is needed is not a new forming of the old symbols that have died, but a rebirth of symbols that disclose a new revelation of God, symbols that grow out of, and are therefore understandable, to the "situation of our time." Only if art can rediscover the basic questions to which the Christian symbols are the answers can these traditional symbols come back to life. Or as Mircea Eliade suggests, we must annihilate the symbols and forms (they cannot be repaired) so that we can begin anew. Destruction of form is really an attempt to reconstruct.

Again within these new levels of the relation between art and religion, expressionism is the overriding style. Expressionism, using elements from everyday life, disrupts and thus radically transforms our ordinary experience of reality visually so that the dimension of depth may be revealed. Surface reality is deceptive; only by breaking through the surface do we discover the ultimate reality that is the foundation of being. Breaking through the surface makes religious art possible.

Over the years Tillich continues to use these four categories to describe the relation between religion and art, but instead of levels he describes them as dimensions. Because, as he says, expression and ultimate reality are present in all matters of life and art, religion is not a hierarchical or separate matter. The depth of expression is what concerns Tillich's view of art.

5. Further Nuances of Style. Tillich generates other categories for his analysis of style. He realizes that within expression are nuances of style that must be deciphered. We have already discovered in Tillich's theology of culture a discernment of three stylistic types: form-dominated, import-dominated, and balanced. In 1959, he develops five stylistic elements that are derived from types of religious experience.[10]

They include: (1) numinous-realism, (2) mystical, (3) critical-realism, (4) prophetic-critical, and (5) expressionism.

1. Numinous-realism depicts ordinary objects of reality in a distorted form that is strange, mysterious, and ambiguous so that the holy may be disclosed. Primitive art, Braque, de Chirico, and Chagall all present the element of numinous realism in their work. It is based on the sacramental experience of reality; all of encountered reality can bear the holy. The potential danger is idolatry. In the case of art, the symbolism can lose its power and become artificial. The symbol would no longer open the depths where the holy may be revealed; instead, the symbol would become the object of worship.

2. Second is the Mystical type which grasps for ultimate reality without the mediation of things. This type can develop an equation of God with nature. In art, particularity gives way to a visual continuum that contains the potentialities of the particular. Chinese landscapes, Impressionism, and non-representational art contain this stylistic element. Ultimate reality is expressed through the basic structural elements of reality. Its danger lies in an artistic emptiness that is void of sacredness. Nothing is expressed.

3. The Prophetic-protesting type of religious experience is critical of the sacramental system, since it subjects nature to human will. Ultimate reality is manifested in history. Critical-realism is the corresponding stylistic element in art. It concentrates on surface reality, for it seems to hide ultimate reality more than express it. It can express ultimate reality, because it makes us look at the familiar in a new way: our callousness to daily life is shattered. Bosch, Goya, and Beckmann (1884-1950) employ critical-realism as they reveal the injustices of the world. A fascination with the ugly that is void of artistic form but instead presents distorted reality is the danger of this element. We are presented with the negativity of life without hope.

4. Religious humanism that sees God in humanity and humanity in God produces the prophetic-critical stylistic element. Future perfection is anticipated in this idealistic type. Tillich reveals a new openness to

idealism when he states that he shared the mood of despising idealism and its inability to mediate ultimate reality until he realized that idealism anticipates the highest possibilities of being. Idealism presents paradise lost and hopes for paradise regained. The essential, undistorted character of humanity is revealed in idealism.

The danger posed by such a humanism is confounding the superficial, sentimental beautifying realism and idealism. Beautifying realism is the result of an unholy marriage of realism with idealism: natural form is depicted with dishonest, idealizing additions. The result is kitsch. Idealism reveals the potentialities in the depths of reality. Poussin (1594-1665), Ingres (1780-1867), Picasso's blue period (1901-1904), and Rousseau (1844-1910) exemplify anticipation of the ideal; they are not kitsch.

This change in Tillich's thought regarding idealism partially resolves the problem regarding his aesthetic experience with Botticelli's *Madonna with Singing Angels* and his disregard for Renaissance art. His change of mood reflects an openness to the revelatory character of idealism. The perfection of form in Botticelli's painting expresses ultimate reality because it represents the essential being undistorted.

5. The fifth and last stylistic element is based on the ecstatic-spiritual religious type of experience. It is specifically the expressionistic movement. Ultimate reality is powerfully manifested when this stylistic element disrupts form to create new form. The Gothic, Van Gogh, Munch, Marc (1880-1916), Nolde (1867-1956), and most modern art since Cézanne present, and are dominated by, the expressionistic element. A potential danger exists even in Tillich's favored style, expressionism: the subjectivity of the artist can predominate so that the artistic form is arbitrary and shallow.

Tillich concludes with the following definition of religious art: "If art expresses reality in images and religion expresses ultimate reality in symbols then religious art expresses religious symbols in artistic images."[11] We have seen that form can swallow import in the form-dominated styles and that religious sentimentality can destroy artistic form. These are shortcomings in the area of religious art. Tillich has remained consistent throughout his career that the expressionistic element has the strongest affinity to religious art. It breaks through the accepting attitude of naturalism and the anticipation of idealism to reach into the depths of ultimate reality.

Tillich struggles with art and its relationship to religion. His distinctions are helpful because they provide something concrete for the viewer to grasp and appraise. Tillich must be applauded for taking art seriously when no other theologians during his time were. His unique and consistent interest in the visual arts opened the door for subsequent theologians, such as David Tracy and John Dillenberger, to look at art seriously. Tillich revealed to the theological world that art is theological in its own right. Yet he is limited in his scope. He allowed himself to be caught in a favorite style and let all other styles of artistic expression revolve around it.

Tillich did not live to witness the post-modern art which we are presently encountering in the works of artists such as George Segal (1924-), Stephen de Staebler (1933-), and Christo (1935-). They attempt to reinvigorate representational art in a new way, suggesting an answer to Tillich's concern for reclaiming traditional symbols.

We have seen in the evolution of Tillich's thought on art that expressionism reveals the breaking down of idealism and naturalism. Expressionism searches for the new and its depths by disrupting what is. The religious element manifested in expressionism unifies subjective and objective reality, for it does not identify itself with objective naturalism or subjective idealism. Expressionism is for Tillich the quintessence of a religious style. It is a particular style that reveals the religious dimensions of life because it discloses the human condition in all of its anxiety. The following passage from *The Courage to Be* summarizes Tillich's response to Expressionism.

> The combination of the experience of meaninglessness and of the courage to be as oneself is the key to the development of visual art since the turn of the century. In expressionism ... the surface structures of reality are

> disrupted. The categories which constitute ordinary experience have lost their power. The category of substance is lost: solid objects are twisted like ropes; the causal interdependence of things is disregarded;...the spatial dimensions are reduced or dissolved into an horrifying infinity. The organic structures of life are cut to pieces which are arbitrarily recomposed: limbs are dispersed, colors are separated from their natural carriers...The world of anxiety is a world in which the categories, the structures of reality, have lost their validity.[12]

Tillich was greatly moved by the style of Expressionism, and he makes it the standard by which to measure all other art. The result is an aesthetics that is limited and problematic yet insightful and encouraging.

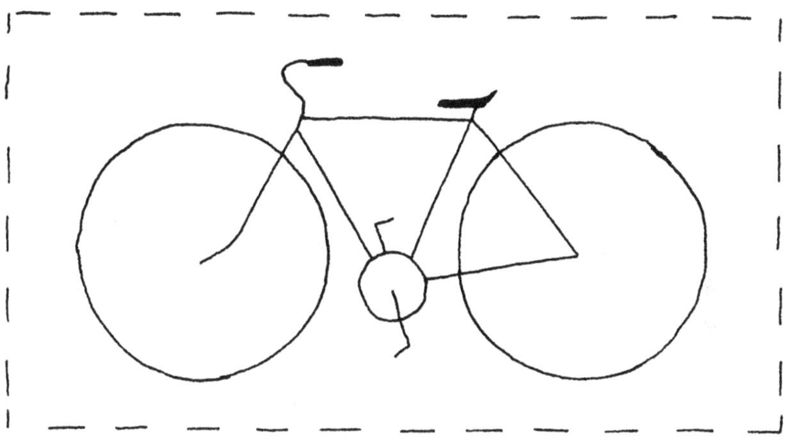

NOTES

1. See Paul Tillich, "Art and Society" in *OAA*, p. 35.

2. Paul Tillich, *Systematic Theology*, Vol. 1 (Chicago: University of Chicago Press, 1951), p. 240.

3. Paul Tillich, "One Moment of Beauty" in *OAA*, p. 235.

4. Paul Tillich, "Existentialist Aspects of Modern Art" (1955) in *OAA*, p. 93.

5. Tillich, "Art and Society," p. 36.

6. Ibid.

7. Paul Tillich, "Existentialist Aspects of Modern Art" in *OAA*, p. 95.

8. Ibid.

9. Thomas Merton, *Disputed Questions* (New York: Harvest/Harcourt, Brace, and Jovanovich, 1985).

10. See Paul Tillich, "Art and Ultimate Reality," (1959) in *OAA*, pp. 139-157.

11. Ibid., p. 151.

12. Paul Tillich, *The Courage to Be* (New Haven: Yale University Press, 1952), p. 146.

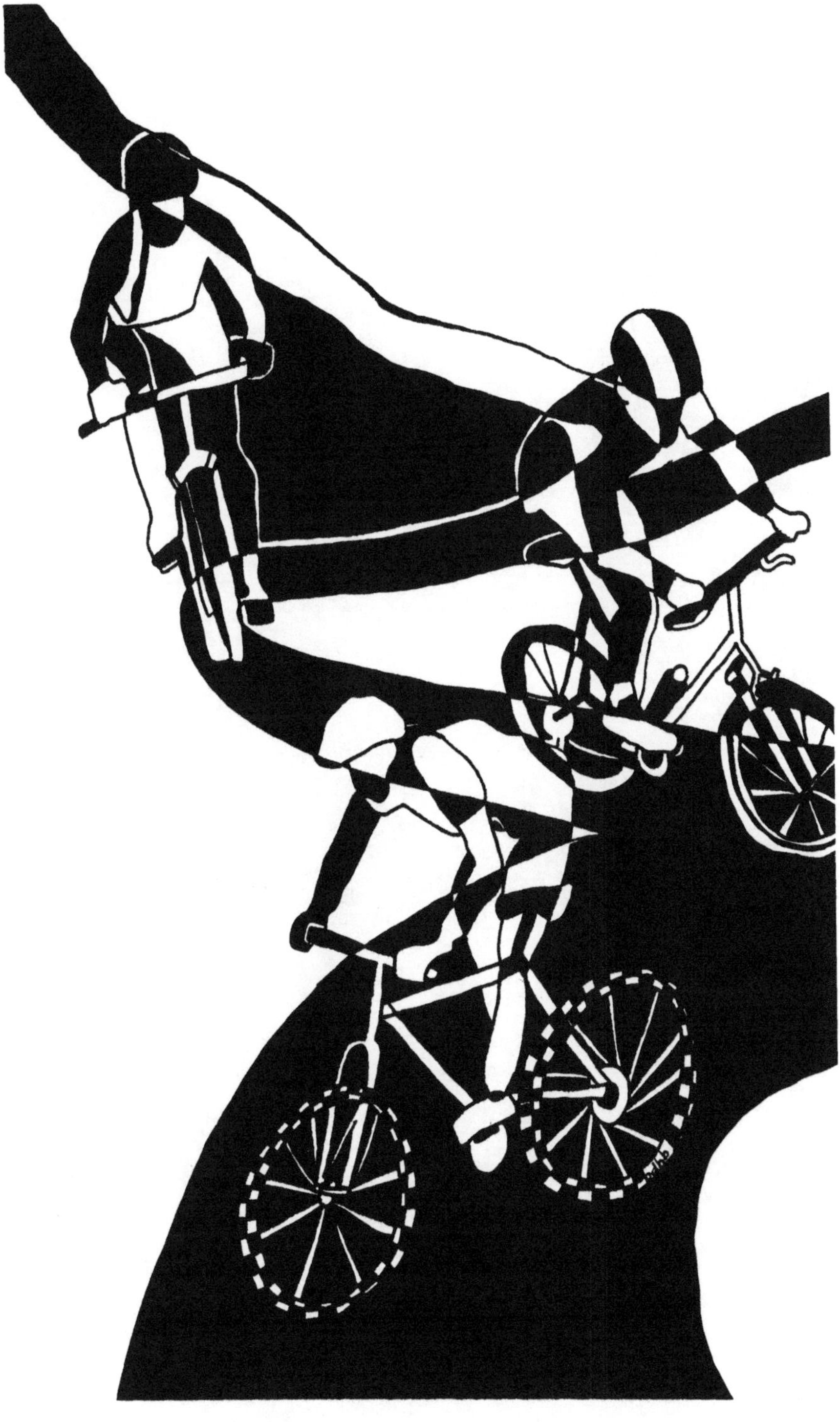

Chapter Ten

Tillich's Theology of Art Re-Examined

from a Polanyian Perspective

In the previous chapters, I have analyzed and extended Michael Polanyi's thought as it bears on art, summarized Paul Tillich's aesthetics, and considered each thinker separately so that my treatment of their thought is independent. In this final chapter, I use the aesthetics I have constructed based on Polanyi's epistemology to examine Paul Tillich's ideas about art and its relation to religion. This process enables us to appreciate and appropriate Tillich's theology of art, to perceive the limitations of his categorizations, and to move beyond the problems and limitations that Tillich's approach poses.

We begin with a description of the limitations Tillich's aesthetics poses when he applies criteria of aesthetic value to art so that he can discern art's relation to religion. Next, we discover how a synthesis with Polanyi's epistemology allows Tillich's theology of art to clarify the relation between art and religion. Finally, a Tillichian description of the task of the art theologian is found significant when enhanced by this Polanyian aesthetics. Within each of the areas delineated, Tillich's limitations and contributions are considered. Let us begin by comparing Polanyi's and Tillich's thought on religion in general.

Paul Tillich and Michael Polanyi are not strangers that I have chosen to introduce to one another. On 21 February 1963, Polanyi and Tillich met, during the Pacific School of Religion's Earl Lectures, for an informal public conversation. In that conversation, Polanyi criticized Tillich for separating the objectivity and detachment of science from all other human concern. In Polanyi's view, all reality that we know, including things in science, are known through our powers of active participation. The knower interiorizes the particulars of an entity and relies on them subsidiarily so as to attend to the whole that is formed by them. S/he dwells in particulars to grasp the joint meaning of a comprehensive entity. This is the structure of meaning in all cases.

Polanyi, like Tillich, emphasizes existential, unresolvable tensions in his theory of knowledge. An assertion of knowledge is based on our belief in its truth, yet it includes an uncertainty due to our inability to know reality completely. We can never fully know in what new way reality may manifest itself in the future. Polanyi regards the Pauline scheme of faith, works, and grace as "the only adequate conception of scientific discovery. The discoverer works in the belief that his [*sic*] labors will prepare his mind for receiving a truth from sources over which he has no control."[1]

Knowing is a personal striving. The knower strives passionately to understand by entrusting his/her self to clues that intimate meaning. Knowledge, once grasped, points beyond itself to indeterminate future manifestations. Knowledge of natural reality may extend continuously into knowledge of supra-natural reality then Supra-Natural reality. Successively more complex integrations of particulars are indwelled to reveal deeper coherences of meaning. Such is the path from scientific knowledge to religious knowing.

That religious knowledge is related to non-religious knowledge is confirmed for Polanyi in the thought of Tillich.[2] Religion is an operative theme that relies on ordinary experience for its basis. Religion employs the raw material of experience in the building up of its own universe. Everyday experience makes religion possible. But religion does not propose assertions of fact; thus, religion is not subject to the conditions of natural reality. Religion is an articulate system that is made possible but not explicated by the everyday. Ordinary experience is subject to but not dependent on the transforming intent of religion—religion re-casts non-religious experience into its own context of religious meaning.

Likewise, historical and scientific knowledge provide unique contexts of meaning that may draw upon or else be at odds with religious meaning. Religion, though, is not subject to scientific and historical analysis. Religion conveys an internal thought that can only be validated through personal surrender into its unique integrative framework. Polanyi aptly makes the point that testing religious claims through natural means is futile. While he refers to the supernatural in the following quotation, the reader should understand his use of the "supernatural" in terms of the "supra-natural" used throughout this study.

> It is illogical to attempt the proof of the supernatural by natural tests, for these can only establish the natural aspects of an event and can never represent it as supernatural. Observation may supply us with rich clues for our belief in God; but any scientifically convincing observation of God would turn religious worship into an idolatrous adoration of a mere object, or natural person.[3]

Scientific and historical criticism may account for the failure of religion to inspire a people to recognize it as a legitimate dwelling place, but science and the social sciences can never adequately account for religious meaning.

Religion provides a unified system of meaning. It is a comprehensive integration that can use all experiences to convey meaning. Religion may also serve other intellectual systems as their theme. Religion relies on, but is not contingent upon, natural experience. Religion and organic experience are mutually related as are all other systems of articulate frameworks. All knowing is interwoven, since ideas draw on each other while attending to realities which are unique to each respective framework. Systems of knowing, including religion, are pregnant with as yet-unknown discoveries. Knowing is progressive because it thrusts toward new manifestations of reality.

The above synthesis draws on the ideas of Polanyi and Tillich where they intersect within the context of religion. Both describe a fiduciary system of religious knowing that is not separate from other systems but continuous with all experience.

Nonetheless, Tillich and Polanyi do not view the relationship between science and the humanities in the same vein. While Polanyi interweaves all types of knowing into one continuous structure, Tillich strictly separates science from the humanities. Since religion is an attitude, according to Tillich, it can penetrate all articulate systems. But science and the humanities are more than attitudes; they are systems of knowing which differ in their epistemological structures. Tillich insists on the independence of the empirical sciences from the cultural sciences. Empirical science is rooted in the ideal of detached, objective knowledge; the cultural sciences, however, involve the creative, social and participatory aspect of knowing. Science, in Tillich's thought, entails a detached mode of certainty that is ultimately verifiable. The personal element is not an active part of knowing; instead, it is to be overcome in the empirical sciences.

Tillich's division between the empirical and cultural sciences gives rise to problems within his aesthetics. He is caught in dichotomous thinking which separates the subject from the object and *theoria* from praxis. Creations of culture fall into the theoretical and practical divisions. I located a debate between Scharlemann and John Luther Adams that tries to determine art's place within Tillich's division between the theoretical and practical. The problem was alleviated by following Adams's lead: art does not fall within the realm of *theoria* or in that of praxis. Art, according to Adams's understanding of Tillich, is on the border.

The problem of designating art's place within the theoretical or practical sphere can be alleviated if the bifurcation between the theoretical and practical is overcome completely. Polanyi's epistemology gives the means for doing just this. Recall that the distinction between the practice of skills (praxis) and the appreciation of the connoisseur (*theoria*) is faint. Theoretical and practical knowledge cannot be formalized but are unspecifiable personal participations in knowing. True theoretical knowledge lies in our ability to apply it practically.

Theoria and praxis are two sides of the same coin. Reality cannot be appreciated, nor received (Tillich), outside of our mindbodies. Only through skillful motoric achievements can theoretical knowledge be exercised and reality grasped. Similarly, the practical achievement of an aim can only be accomplished through our ability to judge and appreciate that aim. We cannot have practical knowledge without theoretical appreciation; we cannot have theoretical knowledge without the exercise of practical skills. Theoretical and practical spheres are unique yet entwined aspects of the act of extending ourselves mindbodily into tacit integrations that achieve an aim and appreciate a comprehensive whole.

Tillich understands *theoria* to be a purely objective reception of reality, while in praxis the knower injects his/her self into reality in order to transform it. In Polanyian personal knowing, the knower does not passively and objectively receive reality; s/he actively intends to understand reality from all that s/he mindbodily is in the world. When the knower apprehends reality, s/he does not intend to transform reality; the knower intends to understand reality and in so doing his/her view of reality is transformed. Not only is the bifurcation between the theoretical and practical overcome in personal knowing but these separate spheres become nonsensical.

The issue of where to place art within Tillich's spheres of *theoria* and praxis is addressed within a Polanyian epistemology of personal knowing. The making and appreciation of art, like all knowing, require an active and passionate submission from ourselves to an independent reality (the artwork). The ensuing knowledge obtained changes the frameworks from which we attend. Knowing is a progressive enterprise that relies on present skills and appreciations that are continuously enlarged and transformed.

1. A Limitation: Tillich's Application of the Criteria of Aesthetic Value

I have expanded Polanyi's criteria of scientific value to the visual arts. The joint application of style, iconography, form, and intrinsic interest is essential to protect the visual arts from the danger of valuing one criterion without the stabilizing influence of the others. The criteria are inseparable and compensate for one another when an artifact is excellent on one criterion but lacking on another.

Tillich's use of the term "content" to refer to subject matter in art is unfortunate and confusing. At this point, I return to the more common use of the term "content" to refer to art's comprehensive meaning. The term "iconography" pertains to an artwork's subject matter.

Tillich sets up a polarity between an artifact's iconography and its import. Since an artwork's meaning results from the mixture of beauty, aesthetic value (style, iconography, form, and intrinsic interest), and originality, import is not in tension with its subject matter. Iconography may be in tension in matters of intrinsic interest. Accordingly, the term "intrinsic interest" rather than "import" will be used. Recall, intrinsic interest refers to a work of art's demand of immediate indwelling, metaphoric quality, and attention to a given field of inquiry. This exchange of terms should clarify my re-examination of Tillich's thought.

Following Tillich's lead, we may state that iconography is meaningful in its expression of form. Iconography is the criterion of least significance, and may even lose its relation to form with an abundance of intrinsic interest. Yet we learn from Polanyi that iconography, too, is an essential criterion in the establishment of aesthetic value. Form, intrinsic interest, and iconography are always in a necessary relationship. Rather than stating that iconography may lose its relationship to form and intrinsic interest, we may more aptly say that while iconography functions jointly with form and intrinsic interest to produce the comprehensive meaning of an artifact's content, the iconography may be low in aesthetic value.

In his book, *Paul Tillich's Philosophy of Art*, Michael Palmer finds difficulty with Tillich's assertion that iconography, form, and intrinsic interest are constitutive of every aesthetic creation. The term "art" in Tillich's scheme designates the presence of iconography, form, and intrinsic interest. According to Palmer, this interdependence removes the crucial margin of uncertainty in art. It is too descriptive and thereby foresees all circumstances and suggests that Tillich's knowledge of art is complete; no new knowledge could modify Tillich's descriptions. Palmer maintains that no necessary relation exists between the criteria of aesthetic value. To assert their joint application is to complete or close the "texture" of art.[4] Art, in Palmer's thought, must be incomplete in its essence; it cannot be completely described.

Palmer is right. Art is an unstable coherence that cannot be entirely delineated; it has an open-texture. Where Palmer errs, like Tillich, is in his presupposition that the recognition of the essential particulars of art (iconography, form, and intrinsic interest) completes art's description. Palmer, in an attempt to correct Tillich's view, separates the criteria of aesthetic value and in so doing claims art to be essentially incomplete. His reductionist theory shies away from establishing any criteria of aesthetic value at all. His assumption is basically the same as Tillich's, that iconography, form, and intrinsic interest wholly describe art.

Tillich bases his analysis of art upon an appraisal of style, specifically: naturalism, idealism, and expressionism. Style, I have contended, is largely unspecifiable. Style suffuses the designated artistic elements of iconography, form, and intrinsic interest, and it integrates the components of an artwork into an harmonious coherence. Style is an underlying unspecifiable element that integrates the three factors of aesthetic value with other elements. Some of these particulars can be designated, such as an artist's socio-historical context, but others cannot. Tillich, in his attempt to describe style, uses descriptions of iconography, form, and intrinsic interest in the name of style. While these are factors of style, they in no way complete the characterization of a given style. Style is more than iconography, form, and intrinsic interest; while these three factors can be described in part, they are by no means fully specifiable in themselves.

Iconography refers to an artwork's representation of natural reality or realism as well as the subject matter depicted. Tillich refers to subject matter in reference to religious or non-religious themes. Subject matter is drawn from the inexhaustible multitude of perceptual objects encountered by humanity; any aspect of natural reality can potentially be integrated into artistic form. The objects selected speak to an artifact's socio-historical context. Yet the simple observation of subject matter is of little aesthetic value in and of itself. Tillich can be appreciated for his insight that an artifact's relation to religion is not dependent on subject matter; a non-religious theme can be deemed more religious than another artwork that depicts religious subject matter.

How an artifact's subject matter is represented is a key element in Tillich's aesthetics. Expressionism, in Tillich's thought, is the quintessence of a religious style, and it relies on the polarity between naturalism and expressionism. Any deviation from naturalism is a movement toward expressionism and religious expressiveness. In Tillich's view, naturalism is a faithful imitation of objects and events from ordinary experience. Religious expression does not rely on representational accuracy; it shatters such accuracy.

Polanyi's epistemology suggests that Tillich's supposed polarity between naturalism and expressionism does not exist. Natural reality is joined with form, intrinsic interest, and other unspecifiable particulars to achieve a deeper coherence of meaning. Data from natural reality is not definitive of style; it is merely a clue to a novel integration. Art is not imitative of nature; it presents its own reality that bears on natural reality. The coherence achieved does not actually exist in nature. The work of art presents a more profound reality that cannot be observed through ordinary perception, nor can it be verbally articulated.

Art's presentation varies in degrees of naturalism and abstraction. Art draws from our experience in the world to present what transcends all natural experience. As it does so, art abstracts various elements from natural reality into an expressive form. The tension in art between degrees of naturalism and abstraction is descriptive, in part, of an artist's aesthetic expression. Unfortunately, Tillich mistook expression for a particular style that could be specified, as exemplified by German Expressionism, instead of a largely unspecifiable quality of art that transcreates and harmonizes its parts.

While Tillich erroneously polarized naturalism and expression, he did grasp the religious quality of expression; it unifies incompatibles into an imaginative coherence that bears supra-natural meaning. Because art harmonizes disparate particulars from natural reality with varying degrees of abstraction, it embodies profound coherences of meaning. I believe this quality that is inherent to all art is its religious expressiveness.

A further problem with Tillich's description of style is that he separates meaning from natural reality. He asserts that empirically known reality is not meaningful in itself even though that reality contains implicit meaning. In Tillich's view, naturalism, in its presentation of what actually is, blocks meaning, since meaning is not innate in the objective structure of reality. Implicit and supranatural (Tillich) meaning must shatter natural reality in order to be expressed. Tillich's view gives rise to two problems. First, he misses the degree of abstraction present in naturalism, and so rejects naturalism's ability to bear supra-natural meaning.

The second problem is that Tillich forgoes the meaning present in natural reality. Tillich's critique of naturalism and idealism (a style he views as potentially dangerous in its deceptive conception of natural reality) is that they do not point beyond themselves. Naturalism and idealism concentrate on explicit reality, not what transcends the surface. Thus, they contradict religious content; they do not make the transcendent existentially present. The transcendent can appear through any form, but naturalism and idealism are derived from finite reality and present themselves as finite where the transcendent is not identified.

While it is true that an accurate rendering loses its aesthetic value when its subsidiary function deflects away from the imaginative form to the natural reality portrayed, Tillich's view suggests that naturalism and idealism have no meaning because they bear directly on an object which is meaningless in itself. They depend on natural reality for their form and thus are meaningless. Herein lies a foundational problem within Tillich's aesthetics.

Polanyi shows clearly that we cannot know any reality apart from meaning. All knowing is meaningful. Natural reality is meaningful, and we understand it through self-centered integrations. It takes on an extraordinary meaning when self-giving integrations are involved. Art, by definition, is self-giving. Its presentation of supra-natural reality is vital to its manifestation. Art incarnates a deeper coherence of meaning than self-centered integrations.

Naturalism and idealism are art forms and thereby present extraordinary meaning. Even if they were self-centered, as Tillich seems to suggest, they would still be meaningful. Their meaning would be of the ordinary sort. The depicted representation would have meaning since it bears on the reality which is its meaning. Art is much more than its treatment of natural form. Tillich's focus on how natural form is depicted leads him to overlook art's comprehensive integration and its supra-natural meaning.

In sum, a peril of valuing one aspect of aesthetic value without the stabilizing influence of the other criteria is evident in Tillich's aesthetics. While Tillich seems to recognize the mutuality of the factors of aesthetic value, his delineation of art hinges on iconography. As a result, he limits art's religious value to one stylistic type (expressionism) and he fails to see the meaningful possibilities naturalism and idealism present.

Like iconography, form concerns itself primarily with observable elements presented in the artifact. In Tillich's view, form accounts for an artwork's degree of religiosity. No equation exists between what is expressed and the means of expression. Form must be transparent to and not identifiable with the ultimate meaning it reveals. Again, it is expressionism's shattering of natural form that presents religious symbolism. Tillich can be affirmed for his observation that there cannot be an equal association between what is expressed and the means of expression.

This observation is applicable to more than religious expression. It is true to semantic relations in general. The from/to structure of knowing advanced by Polanyi demonstrates that verbal meaning is not an equal association of word and object. The word used is not an object interesting in itself. The object that it names is what is interesting as an object. Art extends the from/to structure of semantic relations to include a more complex "into" or self-giving structure. Polanyi's epistemology, which refutes the theory of equal association, suggests that Tillich's statement regarding form is tautological. It is true for communication in general. To state that the dissolution of the equation between what is expressed and its means of expression is the way to reveal religious meaning would be to state that all communication is religious. That assertion could not be defended.

Tillich does not differentiate between iconography and form. When he does consider the formal features of visual art, space, movement, light, and color, it is always in reference to subject matter. He does not conceive of the formal features for their unique qualities but as denotative of representation. He focuses on whether a form is descriptive of natural reality. An art-critical consideration of formal space, for example, may regard the representational quality of space, but it is primarily attentive to the actual size of the artifact, its depth (which need not depend on natural perspective), and the viewer's perceptual relation to the artifact. Tillich does not consider these spacial qualities of form apart from the objects depicted. He evaluates form as if it were the same thing as iconography. Or in the case of the more abstract formal elements, light and color, he is extremely vague, and leaves the reader dissatisfied. The value of art cannot rely on a description of iconography without the comprehensive meaning of the artifact being destroyed and flattened out to a verbal report.

Tillich's enchantment with expressionism is due to (1) its moderate demand of immediate indwelling. High degrees of naturalism or abstraction can inhibit an individual's surrender into an artwork and thereby appear to render it meaningless. Naturalism is less demanding of the imagination since its form is readily identified; abstraction appears uninteresting and incomprehensible in the absence of recognizable form. Natural form abstracted (Tillich's expressionism) is both interesting and immediately demanding of the imagination. A balance between naturalism and abstraction creates a novel coherence that is both familiar and unfamiliar. Even though all styles of art are potentially rich in meaning, we can understand why Expressionism captured Tillich's imagination and guided his aesthetics.

(2) Expressionism exemplifies for Tillich his life experience. It

symbolizes the disruption of his encounter with reality due to the war and the suffering in Germany after the turn of the century. When Tillich readily dwelled in Expressionism, he grasped a more profound meaning of his life experience and thereby discovered the beginnings of his theology of culture.

(3) German Expressionism gave Tillich the material for his creative philosophical expression. With the aid of a book on German Expressionism by Eckart von Sydow, Tillich was able to apply his encounter with Expressionism to a larger comprehensive integration: the theology of culture. Von Sydow helps Tillich understand, then articulate, aspects of this art. Tillich had a powerful experience with the naturalistic painting by Botticelli, *Madonna and Child with Singing Angels.* He describes the event as a never repeated moment of ecstasy in which he encountered the "ground of being." But a cognitive expression of this experience was problematic for Tillich. He could not precisely articulate what he discovered aesthetically because artistic structure transcreates elements found in our world.

While Tillich is carried away and rendered "speechless" by a naturalistic painting, he is moved to draft a theology based on Expressionism. In the process, Tillich dismembers the meaning of the Botticelli painting and of German Expressionism. He gives new meaning to Expressionism through his theology of culture, but the Botticelli meaning is not recovered. The Botticelli experience is too powerful to be explained; it does not fit Tillich's categorical thought. In Tillich's attempt to construct a meaning for the Botticelli, he reduces it to an implicitly religious status. The artifact's own reality is made subservient to a philosophical theory that cannot accommodate an expression of its ecstatic power.

Tillich's articulation of the meaning of Expressionism relies on observable elements that permeate the style. Instead of explaining the meaning of an artwork to which he is attending, he dismembers and renders meaningless artistic elements. He is left with meaningless fragments that different works of art within the style of German

Expressionism appear to have in common. The aesthetic quality destroyed, Tillich re-constructs a new system of meaning to give these particulars significance. He supplies meaning to the specifiable parts, forsaking the comprehensive whole. As the dismembered particulars are integrated into his theory, meaning is discovered in terms of a philosophy so that Tillich's aesthetics supports his theology of culture which in turn was derived from his experience with art.

Unfortunately, Tillich does not recognize the difference between his use of the material of Expressionism for his creative expression in theology and the open texture of the aesthetic experience which is meaningful in itself. He entangled himself in a favorite style and determined, through his theology of culture, that all styles of artistic expression revolve around it. The result is that Tillich makes the descriptive elements of Expressionism *the* measure by which all art and its relation to religion is to be judged.

Tillich's program then follows: art reveals that religion is the substance of culture, while culture is the form of religion. In Polanyian terms, religion harnesses and gives culture meaning, while culture makes religion possible by providing its materiality. Art, then, is meaningful as an entity that embodies religion. The artifact's meaning is determined by religion, and meaningful artistic creativity is sustained by religious intention. Religion is the substance of art.

Tillich's system of meaning presupposes the existence of God and our reliance upon God for meaning. The absence of God assumes the absence of meaning in all human activity. Art is religious by intention. Since all art intends meaning, and meaning relies on the existence of God, all art is religious. Religious (ultimate) meaning is not just one of many ways to interpret art, nor is it a lens through which to interpret art. It is the essence of art.

Tillich does not leave any room for viewers to interpret art from within their unique experience with the artifact. Since meaning does not rely on a particular artwork but instead on a philosophical system, Tillich can do no other than generalize about art. The event of art does not fit the categories of the theology of culture. Thus, particular experience gives way to generalized theory. Tillich does not take into account the tacit location from which each viewer attends to and surrenders into the artwork. His is a closed system of meaning. Nor does he take into account the individuality of artistic style that integrates the artist's skill, location, originality, and all that s/he is mindbodily in the world. Tillich's inconsistent characterization of Giotto's work is a case in point. The integration of the artist's or viewer's belief becomes irrelevant to the aesthetic experience.

Tillich's theology of culture describes specifiable elements that dictate art's meaning. While these elements are integrated into a philosophical system, they are not integrated into particular artworks. Tillich approaches art as a known entity rather than as an invitation to discover the unknown reality that the artifact presents. He omits the quality of art that manifests itself inexhaustibly in the future. He tries to formalize the content of his experience with an artistic style, thereby excluding the indeterminate range of anticipations that may bear on the artifact's reality. We need from Tillich more than an affirmation of art's meaning. We need him to take into account that art must be known in its own terms; that art tells more than we can say.

Polanyi's epistemology helps us recognize that after a few early experiences with art, Tillich focuses on generalized maxims about art derived from his theology of culture. Instead of allowing himself to be carried away by a work of art, Tillich categorizes art. In Tillich's view, we do not begin to understand art's meaning by attending to and surrendering into the artifact, but we rely on particulars of the artifact to confirm our religious framework. Art is not its own reality that reveals supra-natural reality but a subsidiary element of a larger philosophical and theological concern. The result is that Tillich offers systematic generalizations about art that are impersonal and illustrative of its fittingness within the categories of autonomy, heteronomy, and theonomy. His thought lacks the depth and life inherent to art. Rather than deepen a creative encounter with art, Tillich impedes his opportunity to see creatively by focusing instead on an integration of his maxims within the largely unspecifiable skill of art appreciation.

Tillich's constructions may be helpful, however, in the establishment of rules that could be integrated by an inquirer into the field of art and religion. As these maxims are dwelled in, the beholder may attain a greater insight and guidance in art appreciation and theology. The task of Tillich's aesthetics is to excite the viewer with the possibilities of discovering an artifact's theological dimension by guiding the viewer tacitly into an imaginative ferreting out of the artifact's meaning. Tillich's aesthetics may offer clues to an artifact's as yet-unknown meaning; it may point to its theological meaning and even be a part of that meaning, though it does not provide the definitive meaning. At the most, Tillich's analysis of iconography has developed helpful guidelines for a deeper knowledge of art's relation to religion.

2. An Achievement: Three Characteristics of Art in Its Relation to Religion

Polanyi and Tillich imply that art is both creation and discovery. Art discovers supra-natural reality, for it creates novel coherences which embody that discovery. The created artifact in turn invites the beholder to participate in and grasp the discovery. Artistic creativity beckons the artist and viewer alike to discover beyond the natural to the supra-natural. The artifact is an ever-deepening coherence that presents continuous discoveries of meaning as the artist/viewer attends to and surrenders into its comprehensive form. As artistic creativity creates and discovers it does three things according to Tillich: (1) it expresses, (2) it transforms, (3) it anticipates.

1. Tillich's first characterization of art describes art as expressing a reality that transcends the subjective/objective split. We can appreciate this insight in Tillich's recognition of art's presentation of truth. Since Polanyi's epistemology demonstrates that the subject/object bifurcation does not in actuality exist, the assertion that art expresses a subjective/objective harmony points to art's expression of our true

experience of reality. Art embodies the personal extension of ourselves into an independent reality. A synthesis of Polanyi's and Tillich's thought reveals expressiveness to be the spiritual power that harmonizes disparate elements within the artifact. Expressiveness is valued aesthetically but cannot be determined by aesthetic criteria.

2. Art transforms reality to express reality's aesthetic dimension. Art's reality symbolizes the depth-content of natural reality. As a transformer of reality, art is a symbol that presents beauty. Art's beauty is an achievement that transforms natural reality into a symbolic revelation of meaning. We may determine how this transforming characteristic of art is appropriate when its qualities—symbol and beauty—are viewed through Polanyi's epistemology.

The transforming character is dual. First, art transforms the schema or person of the artist and the viewer when they surrender into art, open to its discovery. Second, natural reality is not transformed in its actuality by art, but it is transcreated so that its ordinary meaning is transformed into an extraordinary experience of meaning. Art does not present facts of experience that are transformed; instead, it presents facts of the imagination. These are highly imaginative integrations that do not occur in nature. The reality art presents is not experiential reality transformed; it is its own reality that employs aspects of natural reality subsidiarily. The artist integrates elements that point beyond themselves to a new comprehensive whole. In so doing, art bears on reality and reveals a deeper meaning or supra-natural reality. When the artwork's meaning is grasped, the beholder's view of ordinary reality is transformed.

In Tillich's view, artistic forms are immanent sacramental symbols: they make the transcendent present in their material form. Artistic form is a vehicle for the revelation of the ultimate ground of being that transcends all of life. Art is not autonomous (autobiographical) or heteronomous (dogmatic), but it is theonomous because it bears ultimate meaning. Religious symbols do not refer to the existential worldly realm; they point beyond the tension of the existential polarities to the ultimate level of reality. They are not limited or determined by finitude. Following

Polanyi's lead, we may say that materiality limits but does not determine symbol. Religious meaning embraces all lower levels of reality and gives it Supra-Natural meaning.

Since non-religious art does not exist in Tillich's aesthetics, the distinction of an artifact's religious expression is one of degree, not kind. Inasmuch as all art is religious to a greater or lesser degree, all art is a religious symbol. The degree of an artwork's religious symbolism is based on the same criterion employed to appreciate religious expression: it relies on the tension between naturalism and abstraction. Differences of subject matter and form are not substantive criteria to the perception of a religious symbol. An artistic symbol is most religious when natural form is shattered to allow a greater degree of transcendence to be manifested, that is, naturalism is abstracted.

The symbolic quality of art in Tillich's thought is art's ability to point beyond itself to the dimension of depth. As it does so, art participates in the relationship between Supra-Natural reality and the creator or viewer. This symbolic quality is dependent on our relationship to Supra-Natural reality. When that relationship changes, or takes on new meaning, an artifact once deemed significant will lose its revelatory power.

The symbolic artifact takes on significance as it participates in our relationship with meaning. Tillich assumes this meaning relies on our relationship with God. However, from Polanyi we learn that all knowing is meaningful regardless of a relationship with God. A symbol's meaning is not necessarily religious. We may support Tillich's assertion that an artifact's symbolic meaning and life rely on our participation with the artifact, and the symbol may be deemed religious when we carry into art a religious point of view. Our relationship to Supra-Natural reality is essential to the symbol's religious expression, but not all artistic symbols are essentially religious. Symbolic meaning results from a participatory self-giving integration in which both the symbol and the self, which may or may not include a religious framework, are integrated into that meaning. Like language, symbols are modified over time. A symbol's life changes from one generation or even dies to the next. Our imaginative ability to integrate ourselves into a symbol in light of our intuitive recognition of the symbol's appropriateness bears directly on the symbol's life.

The symbol arises out of what we hold tacitly and is concretized through form so that meaning may be revealed. Since a large part of our tacit knowing is unspecifiable, the symbol cannot be intentionally produced. Its life is subject to the correlation between natural reality and its meaning: natural reality is a vehicle for supra-natural and Supra-Natural reality. This relationship is given form. If we attend focally to the natural form, this relationship is broken, and the symbol no longer participates in the dimension of depth.

Tillich grasps art's basic quality which makes supra-natural meaning visible. He recognizes that a part of the artifact, its iconography, cannot be the object of focus lest its meaning be destroyed. Additionally, the artist's and viewer's framework is lifted up as an important feature in art's ability to reveal meaning. However, a symbol's ability to change us is not identified by Tillich. He does not recognize that art changes the world as we see it; that it calls us into an ever deeper understanding of it.

The heuristic function of art is omitted by Tillich. The knower's beliefs are confirmed and enriched by symbol, but those beliefs are not subject to being changed by the symbol. Tillich is caught between an awareness that art is more than its relation to natural representation and his exemplary use of art. The feature of the symbol's ability to carry the beholder into its own reality and thereby embody the viewer's experience into its life tantalizes Tillich. He is unable to unbind himself into a more open view of art. Being caught in the framework that dichotomizes the subject from the object, he fails to recognize the personal participation that symbol and art require. Tillich proposes that the symbol participates in our life rather than we surrender into and thus participate in the symbol's life. Art is more than symbol. Art's form is interesting in itself (too much so for Tillich!), yet it points metaphorically beyond itself. Art's form must be interesting, yet the metaphorical quality of art requires that the viewer surrender into art openly.

Beauty, for Tillich, is that power which mediates meaning through the transformation of reality into artistic reality. Yet, art does not transform reality but creates its own reality in order to incarnate meaning. Beauty in Polanyi's scheme is much larger than the artistic power that presents meaning. Beauty is what stimulates and attracts our passions because contact with reality has been achieved. It draws us on with a sense of fascination and wonder into discovering ever deepening coherences of reality. This is true for all cases of knowing. Beauty is the pervasive spark that beckons us toward knowledge of all sorts. Beauty belongs to reality's independent existence and its unexpected ways of manifestation.

Art's reality relies on beauty because it draws us into itself. It makes us feel an obligation to surrender into discovering its meaning, and it rewards us by revealing more and unexpected but recognizable meaning. Art expresses beauty, for it makes our inarticulate response to reality visible. In so doing, it presents to the community a vision of reality's deeper meaning. Since beauty resides outside of ourselves, it is the power that mediates meaning. Art, through beauty, offers clues that enchant us into knowing. Since beauty is the attractive aspect of reality, we can appreciate Tillich for grasping beauty's role to mediate meaning as long as it is not limited to artistic reality. Artistic reality makes beauty tangible to the eye so that we may better perceive reality's supra-natural meaning.

In sum, art transcreates natural reality in order to express reality's aesthetic dimension. Art transforms the everyday through its presentation of an extraordinary experience of meaning. The viewer is invited to surrender into a new way of seeing reality. Thus, art via symbol, reveals a deeper meaning in its embodiment of supra-natural reality. The viewer participates in the symbol's life and is enriched and transformed by its event. Beyond symbol and iconography, art is metaphorical. Art creates its own reality which cannot be expressed in another way. When the beholder surrenders into art's reality, s/he experiences beauty which is a lure into deeper meanings of reality. In this way, beauty mediates meaning.

3. Finally, Tillich characterizes art as anticipating salvation or the Kingdom of God. From Polanyi we can better understand this aspect of art as being imaginative. Imagination anticipates as it propels us toward all achievements, be they skills, judgments, or discoveries. Since our understanding of art can never be fully realized, our encounter with art is always one of anticipation. The viewer anticipates the revelation of meaning as s/he attends to and surrenders into the artwork. The continuous imaginative effort reveals deeper meanings the longer we participate in its event. The meaning revealed clarifies our lives, our culture, our world.

As we are able to better understand ourselves, mindbodily beings in the world, we can better grasp our destiny. Our alienation from and our interconnection with one another and God can be made visible to us through art. Art invites communion with our true being, since our selves are embodied into the artifact. Art's ability to make the tacit visible through an imaginative self-giving integration clarifies and deepens our diffuse memories and experiences. Our imaginative participation into works of art liberates us from our existing viewpoints in a radical and unstable way. Drawing upon Tillich's definition of religion, art shakes the foundations of our mindbodily being in the world and builds them up anew.

Art harmonizes incompatibles into a novel meaning which incarnates Tillich's symbol of the Kingdom of God. Our surrender into art enables us to participate in this symbol. In the Kingdom of God, the polarities of life are held through love in a dynamic yet harmonious tension. The harmony embodied in art is not found in nature or human affairs, but its incarnation draws on our unorganized memories, evoking deep emotions in us.[5] The equilibrium of opposed impulses that art presents speaks to our deepest knowing.

We may recognize an essential unity of things as we participate in a work of art. Art embodies a dynamic harmony of extraordinary meaning. Art's presentation correlates with Tillich's symbol of the

Kingdom of God. And we may participate virtually in the Kingdom of God through art. This symbol, which I believe manifested itself to Tillich in his experience with art, is Tillich's greatest contribution toward establishing a theological dimension in art.

Art is an imaginative coherence of incompatibles that requires our surrender into its event. It transcreates our experience of reality from "I-it" into "I-thou" relations. Tillich contributes to this description of art by providing three characteristics which establish art's relationship to religion. Polanyi's thought corrects aspects of these three characteristics. The result is a tripartite characterization of art that bridges art to natural reality and religion. The parts are:

1. Expression: Artistic reality overcomes the subjective/objective bifurcation–it expresses our true experience of reality. This expressiveness is essentially religious since it harmonizes disparate elements from natural reality into an emotionally charged vision, the aesthetic event.

2. Transformation: Art transforms the viewer's ordinary experience by revealing an extraordinary experience of meaning. Via symbol and beauty, art lures the viewer to enter into its event and experience reality more deeply through its presentation of supra-natural reality.

3. Anticipation: Art is an imaginative coherence which propels us continually toward new and deeper meanings. These meanings can clarify our sense of brokenness and our sense of interconnection with reality. Art closes the gap between dimensions of reality we may not otherwise be able to integrate. This anticipation presents dynamic yet harmonious tensions. The beholder of art anticipates and participates in the Kingdom of God, the essential dynamic unity of reality.

Art imaginatively expresses an event which responds to Tillich's symbol of the Kingdom of God. The viewer is invited to participate in that event. As s/he does so, the beholder anticipates deeper coherences of meaning, which when grasped transform the viewer.

3. An Advance: The Task of the Art Theologian

Tillich's theology of culture argues that a theology of art approaches art from the standpoint of meaning. The art theologian has a threefold task: (1) to establish the criteria of aesthetic value; (2) to construct a systematic typology of artistic styles; and, (3) to identify the direction in which a truly religious artistic style may be realized. Tillich informally attempted to accomplish this threefold task in his aesthetics. His aim is an impossible task. He attempts to articulate and categorize style. The result is a narrowing of art into a limited articulate system because the indeterminancies of art are not taken into account. Since style is largely indeterminate, Tillich focuses on a specifiable element of art, subject matter.

The call of the art theologian is to be commended. Yet what Tillich says about art is not as significant to the theological dimension of visual art as is his employment of art to deepen his theology of culture and his Life in the Spirit, especially the Kingdom of God symbol. Polanyi's thought advances the task of the art theologian when applied to Tillich. Through Polanyi we may determine a possible if indefinite task, a theology of art that (1) approaches art from the joint standpoint of aesthetics and theology; (2) recognizes the indeterminate nature of art and the task of the art theologian; and, (3) articulates visual art as theology for the community of believers.

1. The task of aesthetics is to inquire into the relationship of beauty, aesthetic value (style, iconography, form, intrinsic interest), and originality to the content of the artifact. The task of a theology of art is to enlarge the aesthetic inquiry by including criteria of religious merit in an assessment of an artifact's theology. From the standpoint of a religious framework and art appreciation, the art theologian attends to and surrenders into the artifact in order to discover whether a work of art has theological significance. The religious implications the artifact offers may support our existing point of view, but these implications will also change it. Our theology will be enlarged, deepened, or even radically altered in the artistic event. Tillich's neglect of approaching art from the

joint standpoint of a religious framework and art criticism is his greatest weakness.

2. The discernment of an artwork's theology is largely indeterminate, especially in three ways. (1) The art theologian relies on an unstable entity of meaning (the artifact) in which the clues to that meaning are largely unspecifiable. (2) The imaginative effort required to attend to and surrender into the artwork is imbued with integrative principles which are undefinable. (3) The artwork's reality presents a meaning which is inexhaustible.[6] Attempts to formalize the artifact's theological meaning limit the artwork and replace it with assertions that no longer rely on the artifact. Rather than attending to and surrendering into the artifact focally, the artwork becomes a subsidiary clue to a theological formulations. This is a risk the art theologian enters whenever s/he endeavors to reflect upon the artwork's theological dimension.

The theology discovered can be articulated in part, but the art theologian cannot exhaustively define the artifact's meaning. Aesthetic discovery can never be wholly described. It can be transmitted only by the viewer's subsequent surrender into the artwork while dwelling in the clues acquired from the art theologian. The grasping of an artwork's meaning involves knowledge of which we are unable to tell. The theological dimension in art is more truly embodied in the work of art than it could ever be expressed in prose.

The art theologian's second task, then, is indeterminate and dual in its aim. In an act of personal judgment, the art theologian is called to articulate the artifact's theological meaning, while maintaining an awareness of the indeterminacy such assertions contain. As the art theologian attends to and surrenders into the artifact, s/he keeps his/her imagination fixed on the artifact's growing meaning in hopes of realizing the theological dimension. In the process, s/he needs to be open to the artwork's intrinsic intelligibility which may or may not be theological. A theological analysis of an artwork, which destroys its comprehensive meaning, must always be countered with a return to the artifact. The reader/viewer is reminded that the theological analysis presented is partial and can enlarge and deepen the artwork's meaning if and when the beholder surrenders into the actual artifact.

While Tillich, our role model, did not candidly allude to the indeterminate nature of his aesthetic, and typologies in general, he did so tacitly. The way his ideas changed over the years (his new appreciation for idealism, his movement from typing art in levels to dimensions, his typological nuances based on religious experience rather than his theology of culture), his inconsistencies (Giotto, for example), and his theology that claims all we do is fragmentary and ambiguous, suggest that he recognized the indeterminate nature of art and the task of the art theologian. Our understanding of Tillich, the art theologian, necessitates that we participate in the very works of art which inspired Tillich. Studies in Tillich's thought should include trips to an art museum.

3. The art theologian articulates the meaning s/he grasped when participating in the artifact's event and shares this new vision with the community of believers. Its validation is then subject to the community's ability to interiorize the theological assertions into a new subsidiary participation that enriches the artifact's meaning, religious meaning, or both.

In the course of discerning the theological dimension of art, the community of interpreters exercise personal judgment and a creative surrender to and into the artifact in order to understand, apply, and break out from maxims, tradition, and frameworks that are indwelled. Otherwise, theological aesthetics, such as Tillich's, fall into generalities or authoritarianism that impede the imagination; a theology of art, then, becomes an empty shell. When the assertions and traditions of an art theology, including Tillich's aesthetics, are integrated into a skillful performance of art appreciation, the art theologian's claims become a living part of the tradition that views art theologically. A theology of art is progressive when it thrusts its viewers toward new manifestations of an artifact's meaning.

Like all knowing selves, the art theologian is involved in a process of building up and innovating. Novel artistic integrations break the art theologian out from theories which can be binding, while religious and aesthetic values offer stability and continuity to an unstable coherence of meaning. As the art theologian submits into the artwork, s/he is required to break out from existing frameworks into a new theological meaning which reorganizes his/her theological and aesthetic schema. The art theologian is never satisfied to rest in meaning grasped. The call to discover deeper theological meanings through and in art is ever sequential and never ending.

Studies in the thought of Paul Tillich are concerned with those from the community of theologians and philosophers that influenced his thought. Tillich's theology clearly articulates a creative theology derived, in part, from his experiences with art. It is essential that at least the following are included as formative of Tillich's thought: Botticelli, Giotto, Grünewald, Nolde, Cézanne, Picasso.

The art theologian is concerned with a new creation and relies on tradition, criteria of aesthetic and religious merit, the opinion of the artistic and religious communities, and a belief that art bears theological meaning. Even though the artifact may require a revision of traditional aesthetic and religious values, the art theologian will still apply those values and tradition to the joint meaning of the artwork. They are the common, albeit partial, ground upon which visual art may be claimed as theology.

The art theologian must try to avoid suppressing by mistake an artistic discovery that bears on theology, the presentation of which may at first appear meaningless on account of novelty or the excellence of one criterion to the seeming lack of another. The art theologian and his/her community are called to share their experience with art to inform their tradition and religious and aesthetic values. They must give themselves with committed indwelling to a task that is fragmentary and ambiguous. Viewing art theologically is an imaginative heuristic act that requires surrender into its event, is indeterminate and unstable, and grounded in communal acceptance.

NOTES

1. Michael Polanyi, "Faith and Reason," *Journal of Religion*, 41 (October 1961), p. 247.

2. Michael Polanyi, *PK*, p. 283, note 1.

3. Ibid., p. 284.

4. See Michael F. Palmer, *Paul Tillich's Philosophy of Art* (Berlin and New York: De Gruyter, 1984), pp. 181-2.

5. Michael Polanyi, WP, p. 669.

6. See Michael Polanyi, CI, p. 106.

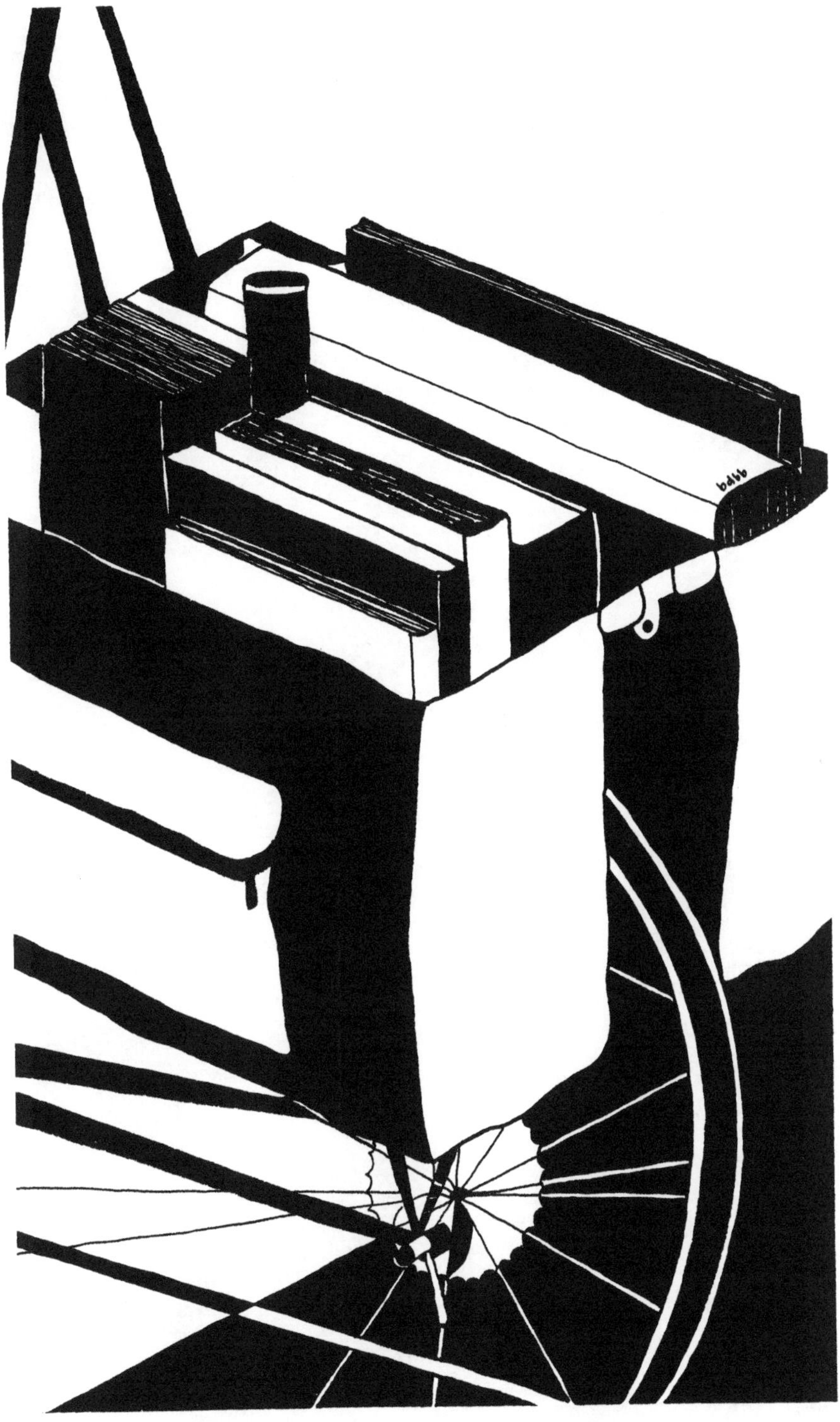

BIBLIOGRAPHY

1. On Polanyi

A. Primary Sources

Polanyi, Michael. *Beyond Nihilism*. Cambridge, England: Cambridge University Press, 1960.

____. *The Contempt of Freedom*. 1940; reprint, New York: Arno Press, 1975.

____. "The Creative Imagination," *Chemical and Engineering News* 44 (25 April 1966): 85-93; reprint Denis Dutton and Michael Krauz, eds. *The Concept of Creativity in Science and Art*. The Hague: Martinus Nijhoff Publishers, 1981: 91-108.

____. "Faith and Reason," *Journal of Religion* 41 (October 1961): 237-247.

____. *Knowing and Being*. Marjorie Grene, ed. Chicago: University of Chicago Press, Phoenix Books, 1969.

____. *The Logic of Liberty*. London: Routledge & Kegan Paul, Limited, 1959.

____. "Meaning: A Project," unpublished lectures forming background of *Meaning*. University of Chicago; University of Texas, Austin; February, March, 1969.

____ and Harry Prosch. *Meaning*. Chicago: University of Chicago Press, 1975.

____. "Points from a Conversation with Paul Tillich." In *Gelwick Microfilm Collection of the Non-Scientific Writings of Michael Polanyi*. Pacific School of Religion, Berkeley, Cal.; 21 February 1963.

____. "Problem Solving," *The British Journal for the Philosophy of Science* 8 (August 1957): 89-103.

____. *Personal Knowledge*. Chicago: University of Chicago Press, 1958; corrected ed., 1962.

____. *Science, Faith, and Society*. Reprint of 1946 London edition. Chicago: University of Chicago Press, 1964.

____. "Skills and Connoisseurship," *Atti Del Congresse di Metodologia* Torino (17-20 December 1952).

____. *The Study of Man*. North Staffordshire: University College, 1959; reprint Chicago: University of Chicago Press, 1963.

____. *The Tacit Dimension*. New York: Doubleday, 1966; reprint Gloucester, Mass.: Peter Smith, 1983.

____. "A Theory of Poetry." Unpublished letter. In *Gelwick Microfilm Collection of the Non-Scientific Writings of Michael Polanyi*. Pacific School of Religion, Berkeley, Cal.; 19 May 1953.

____. "What Is a Painting?" *American Scholar* 39 (Autumn 1970): 665-669.

B. Secondary Sources

Adams, Doug. *Transcendence with the Human Body in Art: Cristo, Stephen DeStaebler, Jasper Johns, George Segal*. New York: Crossroad, 1991.

____ and Phil Mullins. "Conscience, Tacit Knowledge, and the Art of Judgement: Implications of Polanyi's Thought for Moral Reflection." *Soundings* 66:1 (Spring 1983): 34-45.

____ and Phil Mullins. "Meaning with the Arts: The Implications of Polanyi's Epistemology for the Arts." *Studia Mystica* 1:2 (Summer 1978): 28-48.

Apczynski, John V. *Doers of the Word: Toward a Foundational Theology Based on the Thought of Michael Polanyi*. Missoula, Montana: Scholars Press for the American Academy of Religion, 1977.

____. "Truth in Religion: A Polanyian Appraisal of Wolfhart Pannenberg's Theological Program," *Zygon* 17 (March 1982): 49-73.

Gelwick, Richard. "Michael Polanyi: *Credere Aude*: His Theory of Knowledge and Its Implications for Christian Theology." Ph.D. diss., Pacific School of Religion, 1965.

____. "Science and Reality, Religion and God: A Reply to Harry Prosch." *Zygon* 17 (March 1982): 25-40.

____. *The Way of Discovery: An Introduction to the Thought of Michael Polanyi*. New York: Oxford University Press, 1977.

Grene, Marjorie. *The Knower and the Known*. London: Faber and Faber, 1966.

Gulick, Walter. "Michael Polanyi's Theory of Meaning and Reality." *Ultimate Reality and Meaning* (December 1986): 262-268.

Haddox, Bruce. "Questioning Polanyi's *Meaning*: A Response to Ronald Hall," *Zygon* 17 (March 1982): 19-24.

Hall, Ronald. "Michael Polanyi on Art and Religion: Some Critical Reflections on *Meaning*." *Zygon* 17 (March 1982): 9-18.

Iain, Paul. *Knowledge of God: Calvin, Einstein, and Polanyi*. Edinburgh: Scottish Academic Press, 1987.

Irwin, Robert. *Being and Circumstance*. New York: Lapis Press, 1985.

Langford, Thomas A., and William H. Poteat, eds. *Intellect and Hope: Essays in the Thought of Michael Polanyi*. Durham, N.C.: Duke University Press, 1968.

Langford, Thomas A. "Michael Polanyi and the Task of Theology," *Journal of Religion* (January 1966): 45-55.

McCoy, Charles. *When gods Change*. Nashville: Abingdon, 1980.

Milavec, Donald Anthony. "A Theology of Christian Revelation Based upon Michael Polanyi's Interpretation of Scientific Discovery." Ph.D. diss., Graduate Theological Union, 1975.

Mullins, Carl Phillips. "Hermeneutical and Aesthetic Applications of the Thought of Michael Polany." Ph.D. diss., Graduate Theological Union, 1976.

____. "Polanyi's Analysis of Art: A Critical Appraisal." Paper delivered at the Cenntennial Conference on the Thought of Michael Polanyi, Graduate Theological Union, Berkeley, Cal.; 15-16 March 1991.

____. "Religion As an Articulate System." Paper delivered to the American Academy of Religion, Central States Region, 8 April 1991.

Musser, Donald William. "Theological Language and the Epistemology of Michael Polanyi." Ph.D. diss., University of Chicago, 1981.

Poteat, William H. *A Philosophical Daybook: Post-Critical Investigations*. Columbia, Missouri: University of Missouri Press, 1990.

____. *Polanyian Meditations: In Search of a Post-Critical Logic*. Durham, N.C.: Duke University Press, 1985.

Prosch, Harry. "Cooling the Modern Mind: Polanyi's Mission," *Skidmore College Bulletin* 56:4 (August 1971); Saratoga Springs, N.Y.

____. *Michael Polanyi: A Critical Exposition*. Albany, N.Y.: State University of New York Press, 1986.

____. "Polanyi's View of Religion in *Personal Knowledge*: A Response to Richard Gelwick," *Zygon* 17 (March 1982): 41-48.

Reist, Benjamin A. "Faith and the Natural Sciences: Focus on the Thought of Michael Polanyi," *Pacific Theological Review* (Spring 1983); San Francisco Theological Seminary.

Sanders, Andy F. *Michael Polanyi's Post-Critical Epistemology: A Reconstruction of Some Aspects of "Tacit Knowing"*. Amsterdam: Rodopi, 1988.

Scott, Lady Drusilla. *Everyman Revived: The Common Sense of Michael Polanyi*. Lewes, Sussex: Book Guild, 1985.

Thompson, Alexander. *Tradition and Authority in Science and Theology: With Reference to the Thought of Michael Polanyi*. Edinburgh: Scottish Academic Press, 1987.

Torrance, Thomas F., ed. *Belief in Science and the Christian Life: The Relevance of Michael Polanyi's Thought for Christian Faith and Life*. Edinburgh: Handsel Press, 1980.

2. On Tillich

A. Primary Sources

Tillich, Paul. *The Courage to Be*. New Haven: Yale University Press, 1952.

____. *My Travel Diary: 1936; Between Two Worlds*. Jerald C. Brauer, ed. New York: Harper and Row, 1970.

____. *On Art and Architecture*. John and Jane Dillenberger, eds. New York: Crossroad Publishing Co., 1987.

____. *On the Boundary: An Autobiographical Essay*. New York: Charles Scribner's Sons, 1966.

____. *Systematic Theology*. Vol. I. Chicago: University of Chicago Press, 1951.

____. *Systematic Theology*. Vol. II. Chicago: University of Chicago Press, 1957.

____. *Systematic Theology*. Vol. III. Chicago: University of Chicago Press, 1963.

____. *Theology of Culture*. Robert C. Kimball, ed. New York: Oxford University Press, 1959.

____. *Ultimate Concern: Tillich in Dialogue*. D. MacKenzie Brown, ed. New York: Harper and Row, 1965.

____. *What Is Religion?* James Luther Adams, ed. New York: Harper and Row, 1969.

____. "The World Situation." in *The Christian Answer*. Henry P. Van Dusen, ed. London: Nisbet and Company, LTD, 1946.

B. Secondary Sources

Adams, James Luther. *Paul Tillich's Philosophy of Culture, Science, and Religion*. New York: Harper and Row, 1965.

Dixon, John, Jr. "Is Tragedy Essential to Knowing? A Critique of Dr. Tillich's Aesthetic," *Journal of Religion* 43:3 (1963): 271-284.

Dowse, Dimis T. "Tillich's *Analogia Imaginis*: Hermeneutic of Picture and Poetry," *Encounter* (Spring 1986): 109-126.

Gilkey, Langdon Brown. *Gilkey on Tillich*. New York: Crossroad, 1990.

Kegley, Charles W. "Paul Tillich on the Philosophy of Art," *Journal of Aesthetics and Art Criticism* 19:2 (1960): 175-184.

Kegley, Jacquelyn Ann K. *Paul Tillich on Creativity*. Lanham, Md.: University Press of America, 1989.

Mussner, Donald W. "Tillich's Epistemology: An Assessment from a Post-Empiricist Standpoint," *Soundings* (Winter 1986): 423-435.

Palmer, Michael F. *Paul Tillich's Philosophy of Art*. Berlin and New York: De Gruyter, 1984.

Price, Joseph L. "Paul Tillich's Theology of Art." *Soundings* (Winter 1986): 479-498.

Scharlemann, Robert P. "Tillich and the Religious Interpretation of Art." in *The Thought of Paul Tillich, An American Academy of Arts and Science Book*, James Luther Adams, Wilhelm Pauck, and Robert Lincoln Shinn, eds. San Francisco: Harper and Row, 1985

Thompson, Raymond Duane. "Maritain and Tillich: Art and Religion." Ph.D. diss., University of California, Santa Barbara, 1962.

Weisbaker, Dimis Taylor. "The Place of Aesthetics in the Theology of Paul Tillich." Ph.D. diss., Emory University, 1971.

Williamson, Clark M. "The Creative Legacy of Paul Tillich," *Encounter* (Winter 1987): 27-34.

Willimon, William H. "Tillich and Art: Pitfalls of a Theological Dialogue with Art," *Religion in Life* 45:1 (1976): 72-81.

3. On Art and Religion

Adams, Doug. Art Affirms Times to Mourn and Dance," *Church Teachers* (September-October 1988): 58-61.

____ and Diane Apostolos-Cappadona, eds. *Art As Religious Studies*. New York: Crossroad, 1987.

Apostolos-Cappadona, Diane, ed. *Art, Creativity, and the Sacred*. New York: Crossroad, 1983.

Dillenberger, Jane. *Style and Content in Christian Art*. New York: Crossroad, 1986.

____. *Image and Spirit in Sacred and Secular Art*. New York: Crossroad, 1990.

Dillenberger, John. *A Theology of Artistic Sensibilities: The Visual Arts and the Church*. New York: Crossroad, 1986.

Goethals, Gregor T. *The Electronic Golden Calf: Images, Religion, and the Making of Meaning*. Cambridge, Mass.: Cowley Publications, 1990.

Harned, David Baily. *Theology and the Arts*. Philadelphia: The Westminster Press, 1964.

Langer, Susanne K. *Feeling and Form*. New York: Charles Scribner's Sons, 1953.

____. *Philosophy in a New Key*. Cambridge, Mass.: Harvard University Press, 1942.

May, Rollo. *The Courage to Create*. New York: W. W. Norton and Company, 1975.

Merton, Thomas. *Disputed Questions*. New York: Harvest/Harcourt, Brace, and Jovanovich, 1985.

Miles, Margaret. *Carnal Knowing, Female Nakedness, and Religious Meaning in the Christian West*. Boston: Beacon Press, 1989.

____. *Image As Insight, Visual Understanding in Western Christianity and Secular Culture*. Boston: Beacon Press, 1985.

INDEX

Printed by
CPI books GmbH, Leck

FSC
www.fsc.org
MIX
Papier | Fördert
gute Waldnutzung
FSC® C083411

Kolibri 360 GmbH
Ferdinand-Jühlke-Straße 7
99095 Erfurt, Deutschland
produktsicherheit@kolibri360.de